Lecture Notes in Mathematics

Edited by A. Dold and B. Eckmann

T0254714

1326

P.S. Landweber (Ed.)

Elliptic Curves and Modular Forms in Algebraic Topology

Proceedings of a Conference held at the Institute for Advanced Study Princeton, Sept. 15–17, 1986

Springer-Verlag

Berlin Heidelberg New York London Paris Tokyo

Editor

Peter S. Landweber
Department of Mathematics, Rutgers University
New Brunswick, NJ 08903, USA

Mathematics Subject Classification (1980): 11F11, 33A25, 33A45, 55N22, 57S15, 81E99

ISBN 3-540-19490-8 Springer-Verlag Berlin Heidelberg New York
ISBN 0-387-19490-8 Springer-Verlag New York Berlin Heidelberg

Printed in Germany

Printing and binding: Druckhaus Beltz, Hemsbach/Bergstr.
2146/3140-543210

Preface

This volume contains the proceedings of a conference held September 15-17, 1986 at the Institute for Advanced Study in Princeton, New Jersey.

The introductory article provides an account of the recent history of the field of elliptic genera and elliptic cohomology, the central theme of the conference.

The main surprise at the conference was that its original conception was too narrow, and that geometry and physics also enter prominently into this area. For this, see the paper by Ed Witten.

I am grateful to Noriko Yui for permitting her paper on the formal groups of Jacobi quartics, an especially relevant topic for the study of elliptic genera, to be included in this volume.

Thanks are due to David and Gregory Chudnovsky for the suggestion to hold such a conference, and to Bob Stong for substantial advice throughout. It is also a pleasure to thank the School of Mathematics at the Institute for Advanced study, for providing the setting for the conference, and especially Linda Sheldon for much aid. Partial financial support was provided by the National Science Foundation.

Conference Talks

S. Ochanine, Elliptic genera for S^1 manifolds

P. Landweber, Periodic cohomology theories defined by elliptic curves

D. Chudnovsky and G. Chudnovsky, Elliptic formal groups over $\mathbb{Z}$ and $\mathbb{F}_p$ in applications to topology, number theory and computer science

R. Stong, Dirichlet series and homology theories

D. Ravenel, BP-theory for number theorists

M. Hopkins, Characters and generalized cohomology

J. Morava, The Weil group as automorphisms of the extraordinary K-theories

D. Zagier, Modular forms, elliptic functions, Jacobi forms

E. Witten, Elliptic genera and quantum field theory

J. Lepowsky, Infinite dimentional algebras and modular functions

J. Stasheff, Homotopical Lie representations in theoretical physics

Table of Contents

ELLIPTIC GENERA: AN INTRODUCTORY OVERVIEW

by Peter S. Landweber
Rutgers University, New Brunswick, N.J. 08903
Institute for Advanced Study, Princeton, N.J. 08540

The aim of this article is to give an account of the development of the field of elliptic genera. Much of the interest since the fall of 1986 has concerned geometry and physics, but the emphasis here will be on the growth of this area between the fall of 1983 and the fall of 1986, for the benefit of those attracted to the field and wanting to know something of its origins.

At the outset, let me point out that my work in this are has been joint with Bob Stong. I would also like to thank Don Zagier for suggesting the usefulness of such an introductory article.

Prehistory. Here are some of the older results that have played major roles in recent developments.

A. Let the circle group S^1 act smoothly and nontrivially on a closed connected spin manifold M^{2n}. Then one has the Dirac operator [4]

$$D:\Gamma(S^+) \longrightarrow \Gamma(S^-)$$

on spinor fields, for which

$$\mathrm{index}(D) = \hat{A}(M^{2n})$$

where

$$\hat{A}(M) = \hat{A}(M)[M], \ \hat{A}(M) = \prod_i \frac{u_i/2}{\sinh(u_i/2)}$$

if M has total Pontrjagin class

$$p(M) = \prod_i (1+u_i^2).$$

Atiyah and Hirzebruch showed in 1970 [3] that the existence of such an S^1-action implies that $\hat{A}(M) = 0$. Indeed, in case the S^1-action lifts to the principal spin bundle defining the spin structure, so that one has a refined equivariant index

$$\mathrm{index}^{S^1}(D) \in R(S^1),$$

they proved that this character-valued index vanishes.

B. $\Omega_*^U(X)$ denotes the complex bordism of a space X, a module over the complex bordism ring Ω_*^U. One has the <u>Todd genus</u>

$$\mathrm{Td}:\Omega_*^U \longrightarrow \mathbb{Z},$$

namely,

$$\mathrm{Td}(M) = \prod_i \frac{u_i}{1-e^{-u_i}}[M]$$

if the U-manifold (with a complex structure on its stable tangent bundle) M has total Chern class

$$c(M) = \prod_i (1+u_i).$$

Conner and Floyd [10] showed in 1966 that the homology theory

$$K_*(X) = K_0(X) \oplus K_1(X)$$

dual to complex K-theory can be obtained from complex bordism by tensoring:

$$\Omega_*^U(X) \otimes_{\Omega_*^U} \mathbb{Z}_{\mathrm{Td}} \cong K_*(X),$$

where we view

$$\Omega_*^U(X) = \Omega_{\mathrm{even}}^U(X) \oplus \Omega_{\mathrm{odd}}^U(X)$$

and write $\mathbb{Z}_{\mathrm{Td}}$ to indicate that $\mathbb{Z}$ is made an algebra over Ω_*^U via the Todd genus.

C. Let $R(t) = 1-2\delta t^2 + \epsilon t^4$ with δ, ϵ complex numbers, and suppose that

$$\int_0^x \frac{dt}{\sqrt{R(t)}} + \int_0^y \frac{dt}{\sqrt{R(t)}} = \int_0^{F(x,y)} \frac{dt}{\sqrt{R(t)}} .$$

Thus $F(x,y)$ expresses the addition formula for such an elliptic integral. In 1756, Euler [11] gave the formula

$$F(x,y) = \frac{x\sqrt{R(y)} + y\sqrt{R(x)}}{1-\epsilon x^2y^2} .$$

This formal group is quite beautiful, and is central to the number-theoretic study of elliptic genera.

Turning to more recent times we begin with a problem raised by Ed Witten [26] in October 1983.

0. Again, let S^1 act smoothly on a closed spin manifold M^{2n}, but now consider the twisted Dirac operator

$$D_T : \Gamma(S^+ \otimes T) \longrightarrow \Gamma(S^- \otimes T)$$

for spinor fields with coefficients in the tangent bundle T. Then

$$\text{index } D_T = \hat{A}(M)ch(T)[M],$$

where ch(T) denotes the Chern character of the complexification of T. Assume now that the S^1-action lifts to the principal spin bundle defining the spin structure on M. Then one has a character-valued index

$$\text{index}^{S^1}(D_T) \in R(S^1),$$

and Witten [26, §V] asked if this was in fact <u>constant</u> (as a representation, or character). He found this to hold for actions on homogeneous spaces, and suggested that one might apply bordism techniques to prove it in general.

1. With this question in mind, Lucilia Borsari [6,7] took up the problem of analyzing the bordism of circle actions on spin manifolds. She examined a simplified problem, by dealing with <u>semifree</u> S^1-actions on spin manifolds (actions free on the complement of the fixed point set) and tensoring the bordism groups with the rationals. The most interesting problem raised by this study was to determine the ideal

$$I_* \subset \Omega_*^{SO} \otimes \mathbb{Q}$$

(note that $\Omega_*^{Spin} \otimes \mathbb{Z}[\frac{1}{2}] \longrightarrow \Omega_*^{SO} \otimes \mathbb{Z}[\frac{1}{2}]$ is an isomorphism) generated by spin manifolds admitting semifree S^1-actions of <u>odd</u> type (i.e., the action on each component does <u>not</u> lift to the spin bundle). One sees that I_* is also generated as an ideal by all bordism classes $[\mathbb{C}P(V^{2m})]$, $V^{2m} \longrightarrow B$ being a complex vector bundle of <u>even</u> complex dimension over an oriented base. One sees, further, that both the signature [8] and $\hat{A}$-genus [5] vanish on all $\mathbb{C}P(V^{2m})$'s, a promising prospect.

2. A closer look at Borel and Hirzebruch's work from 1958-59 [5] revealed that on the bundles $\mathbb{C}P(V^{2m})$, with fibres the homogeneous spaces

$$\mathbb{C}P^{2m-1} = U(2m) / U(1) \times U(2m-1),$$

not only $\hat{A}$ but also (using an evident shorthand)

$$\hat{A}\,\mathrm{ch}T, \hat{A}\,\mathrm{ch}(\Lambda^2 T),$$
$$\hat{A}\,\mathrm{ch}(\Lambda^3 T + T \otimes T)$$

and several more characteristic numbers of the form

$$\hat{A}\,\mathrm{ch}(\Lambda^k T + \text{lower terms})$$

vanish [15]. We wanted to understand what was behind this.

3. As to the ideal I_*, one has

$$\Omega_*^{SO} \otimes \mathbb{Q} = \mathbb{Q}[x_4, x_8, x_{12}, x_{16}, \dots]$$

and can choose

$$x_4 = [\mathbb{C}P^2], \quad x_8 = [\mathbb{H}P^2]$$

and $x_{12}, x_{16}, \dots \in I_*$. We conjectured that

$$I_* = (x_{12}, x_{16}, \dots),$$

which was supported by the results given above.

Serge Ochanine [16] proved this equality by introducing the notion of an <u>elliptic genus</u>

$$\varphi: \Omega_*^{SO} \longrightarrow R.$$

This means a ring homomorphism (a multiplicative genus) to a commutative $\mathbb{Q}$-algebra, with $\varphi(1) = 1$, so that its <u>logarithm</u>

$$g(x) = \sum_{n \geq 0} \frac{\varphi(\mathbb{C}P^{2n})}{2n+1} x^{2n+1}$$

is an elliptic integral

$$g(x) = \int_0^x \frac{dt}{\sqrt{R(t)}}$$

with

$$R(t) = 1 - 2\delta t^2 + \epsilon t^4 \quad (\delta, \epsilon \in R).$$

He used residues of elliptic functions to prove that, for φ an elliptic genus, one has

$$\varphi(\mathbb{C}P(V^{2m})) = 0,$$

Since $\varphi(\mathbb{C}P^2) = \delta$ and $\varphi(\mathbb{H}P^2) = \epsilon$ for an elliptic genus, it follows easily that I_* can also be characterized as the elements in $\Omega_*^{SO} \otimes \mathbb{Q}$ killed by all elliptic genera.

4. Returning to item 2, we showed that there is an elliptic genus

$$\rho:\Omega_*^{SO} \longrightarrow \mathbb{Q}[[q]],$$
$$\rho(M) = \sum_{k\geq 0} \rho_k(M)q^k,$$

with the coefficients in this power series of the form given there:

$$\rho_k(M) = \hat{A}(M)\mathrm{ch}\rho_k(T)[M]$$

for suitable virtual bundles

$$\rho_k(T) = (-1)^k\Lambda^k(T) + \text{lower terms}$$

depending on the tangent bundle [15]. Strictly, we saw that $\rho_k(T) \in KO(M)$ for low k, but at first only knew that $\rho_k(T) \in KO(M) \otimes \mathbb{Q}$ in general.

5. We needed number theorists to clarify the situation. We learned from David and Gregory Chudnovsky [9] and Don Zagier [29] that ρ maps to <u>modular forms</u> (see below) for

$$\Gamma_0(2) = \left\{\begin{pmatrix} a & b \\ c & d\end{pmatrix} \in SL_2\mathbb{Z} \,|\, c \text{ even}\right\},$$

$\rho(M^{4n})$ being the q-expansion at ∞ of a modular form of <u>weight</u> 2n. We learned explicit expressions for δ and ϵ, as modular forms of weights 2 and 4, respectively, and that $\rho_k(T) \in KO(M)$ for all k. But we still did not know what was behind this.

To say that f is a modular form for $\Gamma_0(2)$ of weight 2n means that $f:H \longrightarrow \mathbb{C}$ is a holomorphic function on the upper half-plane, for which

$$f\left(\frac{a\tau+b}{c\tau+d}\right) = (c\tau+d)^{2n}f(\tau)$$

for $\begin{pmatrix} a & b \\ c & d\end{pmatrix} \in \Gamma_0(2)$, and with

$$f(\tau) = \sum_{k\geq 0} a_k q^k \qquad (q = e^{2\pi i\tau})$$

and a similar holomorphicity condition at the other cusp $\tau = 0$. It is customary to identify a modular form with its q-expansion at ∞.

6. One sees easily that an elliptic genus always sends Ω_*^{SO} into $\mathbb{Z}[\frac{1}{2}][\delta,\epsilon]$; we shall now take δ and ϵ to be indeterminates of dimensions 4 and 8, respectively, and view $\mathbb{Z}[\frac{1}{2}][\delta,\epsilon]$ as an algebra

over Ω_*^{SO} via the corresponding elliptic genus. We asked if there was an underlying homology theory satisfying a "Conner-Floyd theorem" (see item B). Working with Doug Ravenel [13,14], we sought a homology theory with the homology of a point being

$$\mathbb{Z}[\tfrac{1}{2}][\delta,\epsilon,\Delta^{-1}],$$

where $\Delta = \epsilon(\delta^2-\epsilon)^2$ (the discriminant). Indeed, we proved that

$$X \longrightarrow \Omega_*^{SO}(X) \otimes_{\Omega_*^{SO}} \mathbb{Z}[\tfrac{1}{2}][\delta,\epsilon,\Delta^{-1}]$$

is a homology theory, with periodicity of dimension 24. We could also invert merely ϵ or $\delta^2-\epsilon$, and so get homology theories with periodicity of dimension 8.

Thus, we saw the existence of homology and cohomology theories, related to elliptic curves in the Jacobi quartic form

$$y^2 = 1-2\delta x^2 + \epsilon x^4$$

and to modular forms. These are complex-oriented theories, for which the corresponding formal group is the one found by Euler (see item C). At this point a large number of questions suggested themselves.

7. In September 1986, Ed Witten [27] shed considerable light on the "universal" elliptic genus

$$\rho:\Omega_*^{SO} \longrightarrow \mathbb{Q}[[q]]$$

mentioned above. His prescription, with origins in quantum field theory, is as follows. For an oriented manifold M with tangent bundle T, build

$$S_q(T) = \sum_{n\geq 0} S^n(T)q^n, \quad \Lambda_q(T) = \sum_{n\geq 0} \Lambda^n(T)q^n$$

from the symmetric and exterior powers of T, and then write

$$\begin{aligned} R(T) &= \bigotimes_{\substack{\ell>0 \\ \ell \text{ even}}} S_{q^\ell}(T) \otimes \bigotimes_{\substack{\ell>0 \\ \ell \text{ odd}}} \Lambda_{-q^\ell}(T) \\ &= \sum_{k\geq 0} R_k(T)q^k. \end{aligned}$$

One finds that

$$R_0 = 1$$

$$R_1 = -T$$

$$R_2 = \Lambda^2 T + T$$

$$R_3 = -(\Lambda^3 T + T \otimes T)$$

and that

$$(-1)^k R_k(T) = \Lambda^k T + \text{lower terms}.$$

Then it is easily verified [13,29] that our previous $\rho(M) \in \mathbf{Q}[[q]]$ coincides with

$$\hat{A}(M)\ \mathrm{ch}\left\{\frac{R(T)}{R(1)^{\dim M}}\right\}[M].$$

8. Witten also returned to his original question, about the constancy of the character-valued index of D_T, for an S^1-action on a spin manifold. But now we have all the $R_k(T)$ and $\rho_k(T)$ for $k \geq 0$, and so can ask that

$$\mathrm{index}^{S^1}(D_{\rho_k T}) \in R(S^1)$$

be constant for all $k \geq 0$. I.e., <u>one wants the entire equivariant elliptic genus to be constant</u>.

Serge Ochanine had previously studied the question in this form [17,18] and came close to solving it. As a formality, if a compact Lie group G acts smoothly on a closed oriented manifold M, and if $\varphi: \Omega_*^{SO} \longrightarrow R$ is a multiplicative genus for oriented manifolds, one can define an equivariant extension [17,18]

$$\varphi^G(M) \in \prod_{n \geq 0} H^n(BG;R).$$

One then wants to prove that

$$\varphi^G(M) \in H^0(BG;R),$$

provided that G is <u>connected</u>, M is a <u>spin manifold</u>, and φ is an <u>elliptic genus</u>. It suffices to deal with the case $G = S^1$, which Ochanine did provided the action is semifree [17] or preserves a weakly complex structure [18].

Witten gave arguments in [27] and [28] suggesting that the constancy could be proved in the case of general S^1-actions on spin manifolds. Since his arguments rely on unproved properties of the super-symmetric nonlinear sigma model (or equivalently, of Dirac-like operators on the free loop space $\mathcal{L}M$), the task remained to make his ideas into a rigorous proof.

9. Another very pleasant surprise was to learn that other physicists had independently developed some of these ideas. Although I do not feel competent to review these developments, I do want to cite work by K. Pilch, A. Schellekens and N. Warner ([19,20,21,22]) on anomalies and string theory, leading to modular forms for $SL_2(\mathbb{Z})$ in the case of spin manifolds with $p_1 = 0$; and also by O. Alvarez, T. Killingback, M. Mangano and P. Windey ([1,2,12,25]) on loop space index theorems, in which they give a detailed analysis of the index of the Dirac-Ramond operator, the main theme being to extend to loop spaces the path integral proof of the Atiyah-Singer index theorem.

10. The constancy of equivariant elliptic genera for S^1-actions on spin manifolds has now been proved by Cliff Taubes [24]. He makes Witten's program rigorous. One begins with a spin manifold and views

$$M \subset \mathcal{L}M,$$

$\mathcal{L}M$ being the free loop space and M the constant loops. One wants to generalize the Dirac operator on M to a Dirac-like operator on $\mathcal{L}M$. Taubes finds it sufficient to deal with the normal bundle of M in $\mathcal{L}M$. That is, he deals with "small loops" in M centered above points $p \in M$; a small loop is a map $S^1 \longrightarrow TM_p$ whose Fourier expansion has zero constant term. An important feature is the evident "internal" S^1-action on $\mathcal{L}M$ (in addition to the "geometric" S^1-action arising from an action of S^1 on M), having M as its fixed point set. The rather difficult argument given by Taubes [24] has been simplified in further work of Raoul Bott and Taubes.

11. <u>Prospects</u>. Just as index theory for elliptic operators leads to (and requires) K-theory, one can fairly expect that index theory on $\mathcal{L}M$ will lead to elliptic cohomology. At the moment, the main problems are to give a geometric description of elliptic cohomology, and to clarify the connection with index theory on free loop spaces. One expects a prominent role for loop groups and their representations, and further relations with quantum field theory. One also wants to put elliptic cohomology to further use in the realm of topology.

A superb and timely review of this whole field has been given recently by Graeme Segal [23].

References

1. O. Alvarez, T. Killingback, M. Mangano and P. Windey: String theory and loop space index theorems, Commun. Math. Physics 111 (1987), 1-10.

2. O. Alvarez, T. Killingback, M. Mangano and P. Windey: The Dirac-Ramond operator in string theory and loop space index theorems, Nuclear Physics B (Proc. Suppl.) 1A (1987), 189-216.

3. M. F. Atiyah and F. Hirzebruch: Spin-manifolds and group actions, in Essays on Topology and Related Topics, Springer-Verlag, 1970, 18-28.

4. M. F. Atiyah and I. M. Singer: The index of elliptic operators, III, Annals of Math. 87 (1968), 546-604.

5. A Borel and F. Hirzebruch: Characteristic classes and homogeneous spaces, I,II, Amer. J. Math. 80 (1958), 456-538 and 81 (1959), 315-382.

6. L. D. Borsari: Bordism of semi-free circle actions on Spin manifolds, Rutgers University thesis, 1985.

7. L. D. Borsari: Bordism of semifree circle actions on Spin manifolds, Trans. Amer. Math. Soc. 301 (1987), 479-487.

8. S. S. Chern, F. Hirzebruch and J. P. Serre: On the index of a fibered manifold, Proc. Amer. Math. Soc. 8 (1957), 587-596.

9. D. V. Chudnovsky and G. V. Chudnovsky: Elliptic modular forms and elliptic genera, Topology, to appear.

10. P. E. Conner and E. E. Floyd: The Relation of Cobordism to K-theories, Lecture Notes in Math. 28, Springer-Verlag, 1966.

11. L. Euler: De integrationis aequationis differentialis $m\,dx/\sqrt{1-x^4} = n\,dy/\sqrt{1-y^4}$, Opera omnia XX (1), 58-79, Teubner-Füssli, 1911-1976.

12. T. P. Killingback: World-sheet anomalies and loop geometry. Nuclear Physics B, to appear.

13. P. S. Landweber: Elliptic cohomology and modular forms, in this volume.

14. P. S. Landweber, D. C. Ravenel and R. E. Stong: Periodic cohomology theories defined by elliptic curves, to appear.

15. P. S. Landweber and R. E. Stong: Circle actions on Spin manifolds and characteristic numbers, Topology, to appear.

16. S. Ochanine: Sur les genres multiplicatifs définis par des intégrales elliptiques, Topology 26 (1987), 143-151.

17. S. Ochanine: Genres elliptiques équivariants, in this volume.

18. S. Ochanine: unpublished notes, September 1985; Elliptic genera for S^1-manifolds, conference talk, September 1986.

19. K. Pilch, A. Schellekens and N. Warner: Path integral calculation of string anomalies, Nuclear Physics B287 (1987), 362-380.

20. A. Schellekens and N. Warner: Anomalies and modular invariance in string theory, Phys. Lett. 177B (1986), 317-323.

21. A. Schellekens and N. Warner: Anomaly cancellation and self-dual lattices, Phys. Lett. 181B (1986), 339-343.

22. A. Schellekens and N. Warner: Anomalies, characters and strings, Nucl. Physics B287 (1987), 317-361.

23. G. Segal: Elliptic cohomology, Séminaire Bourbaki, 1987-88, no. 695 (Février 1988).

24. C. H. Taubes: S^1 actions and elliptic genera, Harvard University preprint, 1987.

25. P. Windey: The new loop space index theorems and string theory, Lectures at the XXV Ettore Majorana Summer School for Subnuclear Physics: The Superworld II, Erice, 1987, to appear.

26. E. Witten: Fermion quantum numbers in Kaluza-Klein theory, in Shelter Island, II: Proceedings of the 1983 Shelter Island Conference on Quantum Field Theory and the Fundamental Problems of Physics (ed. R. Jackiw, N. Khuri, S. Weinberg and E. Witten), MIT Press, 1985, 227-277.

27. E. Witten: Elliptic genera and quantum field theory, Commun. Math. Physics 109 (1987), 525-536.

28. E. Witten: The index of the Dirac operator in loop space, in this volume.

29. D. Zagier: Note on the Landweber-Stong elliptic genus, in this volume.

ELLIPTIC FORMAL GROUPS OVER $\mathbb{Z}$ AND $\mathbb{F}_p$ IN APPLICATIONS TO NUMBER THEORY, COMPUTER SCIENCE AND TOPOLOGY.

D. V. Chudnovsky, G. V. Chudnovsky

Department of Mathematics
Columbia University
New York, N. Y. 10027

Introduction.

Formal groups have long been used to solve problems in algebraic geometry, algebraic number theory and topology. In this paper, we describe a few more applications of many concepts borrowed from formal groups. One of them is the relationship between the integrality conditions on power series expansions of functions (representing, say, logarithms of formal groups) and the algebraic and analytic properties of these functions. Among the formal groups that we consider those associated with algebraic curves, particularly with elliptic curves, and, in general, with laws of addition on Abelian varieties, play the most important role. In addition to topology our main areas of focus in this paper are: number-theoretic properties of differential equations (the Grothendieck problem), the uniformization problem and integrality conditions (the Tate conjecture), congruences for coefficients of algebraic differentials (related to Schur congruences), and a variety of applications of interpolation on algebraic curves over finite fields.

This work started a few years ago when we got interested in formal groups in connection with the Grothendieck conjecture (see §1). Our interest in formal groups and their application grew as we met Peter Landweber and got involved by him and Bob Stong in a variety of exciting problems centered around formal groups, characteristic classes and modular forms (see §7). For the last two years, Landweber and Stong conducted an international seminar by correspondence, open to everybody, that generated considerable progress in this field, and this conference is just a physical realization of this virtual seminar. In this paper, which follows our lecture, we will deviate from algebraic topology into computer problems, still firmly holding onto elliptic curves.

The paper is organized as follows. We start in §1 with an examination of the connection between the integrality of coefficients of formal power series expansions and the analytic properties of functions so expanded. Particular attention is devoted to criteria of the algebracity of these functions (the Eisenstein theorem and theorems inverse to it). Subjects include: strict isomorphisms and isogeny of elliptic curves (the elliptic part of the Tate conjecture) following [19]. We devote attention to Honda's and Grothendieck problems on the algebraicity of functions (formal groups) satisfying integrality conditions. In §2 we examine the formal completion of elliptic curves over the fields of positive characteristic, and present new congruences (some of them conjectural) for Legendre polynomials associated with these formal groups. The action of Frobenius on multiplicative, elliptic and general algebraic formal groups over $\mathbf{F}_p$ is studied in §3 from the point of view of primality testing. In §4 we look at the important algorithmic problem of the least complex polynomial multiplication algorithms. We describe popular complexity models and the classical results of Winograd et. al. in the case of infinite ground fields. We have discovered new relationships with linear coding theory and algebraic curves of positive genus over finite fields, which yield improved complexity bounds. In §5 we give an exposition of our theory of interpolation on algebraic curves of positive genus. From this theory we deduce the linear upper bounds on the multiplicative complexity of the polynomial multiplication over finite fields. E.g. the minimal multiplicative complexity $\mu_k(K)$ over the field of scalars k of the multiplication in the extension K of k of degree n is bounded between $3.52 \cdot n$ and $6 \cdot n$ for $k = \mathbf{F}_2$ for sufficiently large n. (Note that for an infinite k, $\mu_k(K) = 2n - 1$.) In §6 multidimensional laws of addition generalizing elliptic ones are displayed in connection with obstruction to the factorization of S-matrices. In §7 we present our formula for the universal elliptic genus in terms of theta functions that reveals its structure as a generating function for modular forms of level two. This formula, which was recently reformulated by Witten from the geometric point of view, was obtained by us in the summer of 1985 as a solution to the integrality problem of Landweber-Stong.

We refer readers for the description of further developments of the elliptic genus study to [64].

We want to thank Peter Landweber for his enthusiasm, encouragement, attention, and support.

We are thankful to R. Jenks and B. Trager for their interest and support.

This work has been supported in part by the N.S.F., U.S. Air Force, and Program O.C.R.E.A.E.

§1. Integrality of coefficients of power series expansions and formal groups defined over $\mathbb{Z}$.

Among power series expansions for number theorists the most interesting are those whose coefficients are (algebraic) integers. Pólya-Szegö [1] call a power series expansion $f(x) = \Sigma_{n=0}^{\infty} a_n x^n$ Eisenstein, if all coefficients in the expansion $f(x)$ are rational numbers, and there exists an integer $N \geq 1$ such that all coefficients in the expansion $f(Nx)$ are integral. All algebraic functions over $\mathbb{Q}(x)$, that have power series expansions, are Eisenstein (by the Eisenstein theorem, see [1]). Other power series expansions important in number theory that have (nearly) integral coefficients include modular forms, in particular various combinations of ϑ-constants. Integrality of a power series expansion's coefficients is another way of saying that this expansion is defined mod p for (almost) all p.

In the abstract theory of (one-dimensional, commutative) formal groups the integrality conditions on coefficients of power series expansions produce strong effects. In one example, one looks, following Hazewinkel [2], at a one-dimensional formal group law $F(x,y) = x + y + \ldots \in \mathbb{Q}[[x,y]]$ with the logarithm function

$$\ell(x) = \Sigma_{n=1}^{\infty} \frac{a_n}{n} x^n \in \mathbb{Q}[[x]],$$

i.e.

$$F(x,y) \equiv \ell^{-1}(\ell(x)+\ell(y)).$$

Let $L(s) = \Sigma_{n=1}^{\infty} a_n n^{-s}$ be a Dirichlet series with (rational integral) coefficients a_n, associated with the logarithm $\ell(x)$. Then the power series group law $F(x,y)$ is p-integral (i.e. no denominator of coefficients of the power series $F(x,y)$ is divisible by p) <u>if and only if</u> $L(s)$ has an Euler product for the prime p in the sense that

$$L(s) = \Pi(p,s)^{-1} \cdot L_1(s),$$

$$\Pi(p,s) = 1 + e_1 p^{-s} + e_2 p^{1-2s} + \ldots,$$

$$L_1(s) = \Sigma_{n=1}^{\infty} b_n n^{-s}$$

with $b_n \equiv 0 \bmod p^k$ if $p^k | n$ and $e_i \in \mathbb{Z}_p$.

In particular, if $F(x,y)$ has (rational) integral coefficients, and finite heights mod p for all p, then the corresponding Dirichlet series L(s) has an Euler product for all p with the Euler factors $\Pi(p,s)$ having finitely many terms. These conditions for the existence of the Euler products have an explicit representation as congruences on coefficients a_n mod p^k known as Atkin-Swinnerton-Dyer congruences, see [2]. The Atkin-Swinnerton-Dyer congruences were formulated for formal groups of elliptic curves defined over $\mathbb{Z}$, and proved by Cartier [3], Honda [4,5] (under certain additional conditions) and by Hill [6], Ditters [7,8] and Yui [9].

The crucial property of integrality of F(x,y) led Honda to the following problem: for which Dirichlet series $L(s) = \Sigma_{n=1}^{\infty} a_n n^{-s}$ with integral coefficients, does the corresponding logarithm function

$$\ell(x) = \Sigma_{n=1}^{\infty} \frac{a_n}{n} x^n \in \mathbb{Q}[[x]]$$

define a group law

$$F(x,y) \equiv \ell^{-1}(\ell(x) + \ell(y))$$

over $\mathbb{Z}$ (i.e. $F(x,y) \in \mathbb{Z}[[x,y]]$)? When are these group laws algebraic? See Honda [10].

Honda [10] was particularly interested in logarithms $\ell(x)$ that are defined as "ordinary" functions, particularly as solutions of differential equations (though it is not clear to us, whether he was interested in linear or arbitrary algebraic differential equations). In his work on differential equations with integral power series expansions of solutions, Honda (see his account in [10]) arrived at the Grothendieck conjecture for linear differential equations.

Since we have arrived at linear differential equations, it is time to present another class of functions with nearly integral power series expansion coefficients. This class of functions was introduced by Siegel [11] in 1929 and called a class of G-functions. According to Siegel, a function f(x) with the expansion

$$f(x) = \Sigma_{n=0}^{\infty} a_n x^n$$

is called a G-function if: i) f(x) satisfies a linear differential equation over $\bar{\mathbb{Q}}(x)$; ii) all coefficients a_n are algebraic numbers; and iii) for all n, the sizes $|\overline{a_n}|$ of the algebraic numbers a_n and common denominators $\mathrm{den}\{a_0,\ldots,a_n\}$ are bounded by geometric progressions in n:

$$|\overline{a_n}| \le C_1^n,$$

$$\mathrm{den}\{a_0,\ldots,a_n\} \le C_2^n: n \ge 1$$

with (constants) $C_1 > 1$, $C_2 > 1$ depending only on $f(x)$.

Among G-functions, in addition to algebraic functions, one finds generalized hypergeometric functions ${}_{p+1}F_p\left({a_1,\ldots,a_{p+1} \atop b_1,\ldots,b_p}\middle| x\right)$ with rational a_i, b_j, and various periods of algebraic varieties (solutions of Picard-Fuchs equations) depending on a parameter x. Following the celebrated Siegel's conjecture on the structure of E-functions [Note: the definition of an E-function [11] is identical to the one given above, but $f(x)$ has now the form $f(x) = \Sigma_{n=1}^{\infty} a_n/n!\, x^n$], one can speculate that all G-functions "arise from geometry" (Dwork), i.e. are reducible to integrals of algebraic functions, [12].

Speaking more geometrically, linear differential equations satisfied by G-functions are subject to severe arithmetic constraints.

The "geometric" restrictions imposed on the differential equation by the fact that one of its solutions is a G-function, were subject to much speculation in diophantine approximation theory [12], [13], [14]. These speculations were connected with Siegel's [11] program on the proof of linear independence theorems for values of G-functions at rational (algebraic) points close to the origin, similar to Siegel's famous results on values of E-functions [11]. Early results on G-functions required strong restrictions on differential equations satisfied by G-functions, reflecting conditions of good p-adic convergence of generic solutions of these differential equations [13]. In [15] all restrictions on G-functions necessary to prove the linear independence results were removed. The method used in [15] was the method of Padé approximations by the so-called Germanic polynomials. In [12] we showed that the existence of a nontrivial G-function solution of a linear differential equation implies very good p-adic convergence of generic solutions of this equation. One of our results from [12] can be formulated as

<u>Theorem 1.1</u>: Let $L[y] = 0$ be a linear differential equation of order

n over $\bar{\mathbb{Q}}(x)$ satisfied by a G-function $y(x)$, which does not satisfy any linear differential equation over $\bar{\mathbb{Q}}(x)$ of order $< n$. Then all solutions of the equation $L[y] = 0$ with algebraic initial conditions at an algebraic point $x = x_0$ have G-function expansions at this point.

In particular, a linear differential equation $L[y] = 0$ is globally nilpotent (in the sense of Katz [16]), which, roughly speaking, means that

$$\left(\frac{d}{dx}\right)^{pn} \equiv 0 \bmod(p, L[\cdot])$$

for almost all (density one) p.

The last condition is called the "p-curvature of L is nilpotent" (Katz [17]). A stronger condition,

$$\left(\frac{d}{dx}\right)^{p} \equiv 0 \bmod(p, L[\cdot]),$$

is called "p-curvature is zero", and is the subject of the Grothendieck conjecture, which in fairness, should be called the "Grothendieck-Katz Conjecture".

<u>The Grothendieck Conjecture:</u> Let $L[y] = 0$ be a linear differential equation of order n over $\mathbb{Q}(x)$ (or $K(x)$). If $L[y] \bmod p = 0$ has n linearly independent solutions in $\mathbb{F}_p[x]$ (or in $\bar{K}_{\mathcal{P}}[x]$ for $L[y] \bmod \mathcal{P}$), or, equivalently, $L[y]$ has a "zero p-curvature" for almost all p (almost all prime $\mathcal{P}$ in K), then all solutions of $L[y] = 0$ are algebraic functions.

According to this conjecture [17], strong integrality (Eisenstein-like) properties of power series expansions of all solutions of a given linear differential equation imply that all these solutions are algebraic functions.

Various results of this form would constitute the converse to the Eisenstein theorem. As it turned out, it is possible to prove a variety of positive results whenever the assumptions of near integrality of coefficients of power series expansions--nonarchimedean conditions--are coupled with assumptions on the analytic continuation of an expanded function in the complex plane (its Riemann surface)--an archimedean condition. An early example of such a theorem is the Borel-Polya theorem, according to which power series expansions with integral coefficients, meromorphic in the domain of conformal radius > 1

represent rational functions. Padé approximations methods allowed us to generalize considerably this result to functions uniformized by various classes of meromorphic functions [18]-[19]. Roughly speaking, one of our results [18] means that n functions $f_1(\bar{x}),\dots,f_n(\bar{x})$ in g variables $\bar{x} = (x_1,\dots,x_g)$, $n \geq g+1$, having "nearly integral" power series expansions at $\bar{x} = \bar{0}$, and being uniformized near $\bar{x} = \bar{0}$ by meromorphic functions in $\mathbb{C}^g$ of finite order of growth, are algebraically dependent.

The condition of "near integrality" means that denominators of coefficients in the expansions grow slower than factorials. It seems worthwhile to present one example of our results: Theorem 1.1 proved in our paper [19].

Theorem 1.2: Let $n \geq g+1$ functions $f_1(\bar{x}),\dots,f_n(\bar{x})$ be uniformized near $\bar{x} = \bar{0}$ by n meromorphic functions $U_1(\bar{u}),\dots,U_n(\bar{u})$ in $\mathbb{C}^g$ of finite order $\leq \rho$ of growth under a nonsingular transformation $\bar{x} \rightleftarrows \bar{u}$ (near $\bar{x} = \bar{0}$). Let $f_i(\bar{x}) = \Sigma_{\bar{m}\in\mathbb{Z}^g} a_{\bar{m}}^{(i)}\bar{x}^{\bar{m}}$ be expansions at $\bar{x} = \bar{0}$ such that $a_{\bar{m}}^{(i)} \in K$ for an algebraic field K of degree d over $\mathbb{Q}$. Put $f_1(\bar{x})^{k_1}\dots f_n(\bar{x})^{k_n} = \Sigma_{\bar{m}} a_{\bar{m};\bar{k}}\bar{x}^{\bar{m}}$ for $k_i \geq 0$ and denote

$$\Delta_M = \text{den}\{a_{\bar{m};\bar{k}}: \|\bar{m}\| < M, \|\bar{k}\| < M\};$$

$$\xi = \limsup_{M\to\infty} \frac{\log|\Delta_M|}{M \log M},$$

$$\sigma = \limsup_{M\to\infty}\{\overline{|a_{\bar{m}}^{(i)}|^{1/M}}: \|\bar{m}\| < M\}.$$

Then, if $\sigma < \infty$ and $\xi < \frac{(1-g/n)}{d\rho}$, the functions $f_1(\bar{x}),\dots,f_n(\bar{x})$ are algebraically dependent over K. [Here $\|\bar{m}\| = m_1+\dots+m_g$, etc.].

These results are the converse to the Eisenstein theorem, and can be applied, e.g., to the solution of the Grothendieck conjecture.

For example, we obtained positive solutions to the Grothendieck conjecture, whenever all solutions of a given linear differential equation over $\bar{\mathbb{Q}}(x)$ are uniformized by meromorphic functions of finite order of growth. This is the case of the Lamé equations with integral n (Dwork's problem [20])--see [21]. Other equations for which we have proved the Grothendieck conjecture include all rank one equations over elliptic curves and other curves of positive genus, and all linear differential equations describing the so called finite-band potentials, see [18].

These results on the Grothendieck-Katz conjecture can be applied to a group of interesting questions proposed by Matthews [22]. Matthews formulated his questions in terms of a projective absolutely irreducible variety V over $\mathbb{Q}$ with a (finitely generated)group Γ of automorphisms of $\mathbb{Q}(V)$ and locally trivial (i.e. trivial mod p for almost all p) classes in $H^1(\Gamma,\mathbb{Q}(V)^{\ddagger})$ for the action of Γ on either the additive or the multiplicative group of $\mathbb{Q}(V)$. For most of the cases in [22], Γ corresponded to linear transformations or translations in elliptic curves (tori). The appropriate conjectures reflect the "local-global principle", and are well suited to be studied with our "archimedean + nonarchimedean analyticity conditions". Among Matthews problems were particular cases of the Grothendieck conjecture (solved by us) and the following amusing problem on indefinite integration of algebraic functions. Let Γ be a curve over $\mathbb{Q}$ and D be any derivation of a function field $\mathbb{Q}(\Gamma)$. Let $f \in \mathbb{Q}(\Gamma)$ be such that for almost all p, we can find g in $\mathbb{F}_p(\Gamma \bmod p)$ such that, mod p, $f = Dg$. Is it true then, that

$$f = Dg$$

for some g in $\mathbb{Q}(\Gamma)$? (If an integral is locally algebraic from the same field, is it globally algebraic?) The answer is "yes".

How about other linear differential equations? Most of them are not uniformized by meromorphic functions (e.g. Lamé equations with nonintegral n). Again, the answer seems to depend on the combination of nonarchimedean and archimedean conditions imposed on solutions of a linear differential equation. For example, if solutions of a linear differential equation can be uniformized by an arithmetic group, one can prove a positive answer to the Grothendieck conjecture. Moreover, a much weaker condition: that the monodromy group of a linear differential equation of order n is up to conjugation a subgroup of $GL_n(\bar{\mathbb{Q}})$--implies a positive answer to the Grothendieck conjecture, provided that this linear differential equation satisfies the assumptions of the Grothendieck conjecture [23].

These results on algebraicity properties of functions satisfying archimedean and nonarchimedean analytic continuation properties can be applied to nonlinear differential equations as well. One such application involves formal groups associated with elliptic curves--the subject of this conference. To fix notations, we start with an arbitrary plane cubic model of an elliptic curve E:

$$y^2 + a_1xy + a_3y = x^3 + a_2x^2 + a_4x + a_6.$$

The functions x and y on E are parametrized in terms of Weierstrass elliptic functions:

$$x_E = \wp(u) - (a_1^2+4a_2)/12,$$

$$2y_E = \wp'(u) - a_1x - a_3;$$

$\wp'(u)^2 = 4\wp(u)^3 - g_2\wp(u) - g_3$. An invariant differential on E is $\omega = dx/(2y+a_1x+a_3) = du$. The choice of a local parameter z near the origin is $z = -x/y$ (Tate, [24]). The expansion $\omega = \Sigma_{n=1}^{\infty} b_n z^{n-1} dz$ of ω with $b_1 = 1$, $b_i \in \mathbf{Z}[a_1,\ldots,a_6]$ gives rise to the elliptic logarithm of E:

$$\ell_E(z)\,(=u) = \Sigma_{n=1}^{\infty} \frac{b_n}{n} z^n \quad (= \textstyle\int\omega).$$

If E is defined over $\mathbf{Z}$, i.e. $a_i \in \mathbf{Z}$, then the formal group law $F_E(X,Y)$:

$$F_E(X,Y) = \ell_E^{-1}(\ell_E(X) + \ell_E(Y))$$

is defined over $\mathbf{Z}$.

Honda's [4,5] criterion for isomorphism of one-dimensional group laws over $\mathbf{F}_p$ can be interpreted, in case of elliptic curves, as a nonarchimedean convergence criterion. Combination of this nonarchimedean convergence with the uniformization by the Weierstrass elliptic functions can be used to prove isogeny theorems that are particular cases of the Tate conjecture [19]. For example, let E_1 and E_2 be two elliptic curves defined over $\mathbf{Q}$ that have isomorphic group laws over $\mathbf{F}_p$ (or, equivalently, the same number of solutions mod p) for almost all p. Then E_1 and E_2 are isogenous over $\mathbf{Q}$ [19].

Indeed, according to the Honda criterion [2,4,5] the two formal group laws $F_{E_1}(X,Y)$ and $F_{E_2}(X,Y)$ are strictly isomorphic over $\mathbf{Z}_p$ for almost all p. This means [2] that the power series

$$f(z) \equiv L^{-1}_{E_1}(L_{E_2}(z)) \quad (= z + O(z^2))$$

has p-integral coefficients for almost all p. We can apply now our converse to the Eisenstein theorem (cf. Theorem 1.2 above with $g = 1$, $n = 2$) to two functions:

$$z \text{ and } f(z).$$

These functions have "nearly integral" expansions and are uniformized by the Weierstrass elliptic functions

$$z = z_{E_1}(u), f(z) = z_{E_2}(u),$$

where

$$z_{E_i}(u) = -x_{E_i}(u)/y_{E_i}(u): i = 1,2.$$

Consequently, the combination of nonarchimedean and archimedean conditions of Theorem 2.1 implies that the function f(z) is algebraic over $\mathbb{Q}(z)$, i.e. that the elliptic functions $\wp_1(u)$ and $\wp_2(u)$ are algebraically dependent over $\mathbb{Q}$. This means that E_1 and E_2 are isogeneous over $\mathbb{Q}$. This result is effective, for bounds see [19]. Of course, general results on the Tate conjecture belong to Faltings [25].

§2. Congruences for Legendre polynomials.

Formal groups associated with elliptic curves over $\mathbb{Z}$ that arise as genera corresponding to characteristic classes vanishing on Spin manifolds with odd S^1 action have special integrality properties. As we already remarked, the mere fact that these formal group laws are defined over $\mathbb{Z}$ (the consequence of the algebraicity of this group law and the Eisenstein theorem) implies the existence of the Atkin-Swinnerton-Dyer congruences. These congruences are expressed in terms of Legendre polynomials, and are related to the famous Schur congruences, because appropriate models of elliptic curves for genera of Spin manifolds are the Jacobi quartics, an invariant differential of which has Legendre polynomials as coefficients of power series expansions.

Indeed, an interesting group of congruences, involving values of

Legendre polynomials and Legendre polynomials themselves, are all connected with the action of Frobenius on formal groups arising from Jacobi's quartic

$$(E_\delta): \qquad Y^2 = 1 - 2\delta X^2 + X^4 \overset{\text{def}}{=} R(X)$$

and the formal group law on this elliptic curve, as defined by Euler:

$$F_{E_\delta}(X_1,X_2) = \frac{X_1\sqrt{R(X_2)} + X_2\sqrt{R(X_1)}}{1 - X_1^2X_2^2}.$$

Legendre polynomials $P_n(\delta)$ occur in the expansion of the differential of the first kind on E_δ (one can take this generating function as a definition of Legendre polynomials):

$$\omega \overset{\text{def}}{=} \frac{dX}{Y} = \frac{dX}{\sqrt{R(X)}} = \Sigma_{n=0}^{\infty} P_n(\delta)X^{2n}.$$

Depending on the ring of definition of E_δ, one derives a variety of congruences on $P_n(\delta)$ mod p^k.

Arithmetically the most interesting congruences correspond to the case of an integer $\delta \neq 0,1$, and are Atkin-Swinnerton-Dyer congruences on the coefficients in the formal expansion of the differential of the first kind ω of the elliptic curve (in its minimal model) defined over $\mathbb{Z}$:

$$\omega = \Sigma_{n=1}^{\infty} a_n z^{n-1}dz, \; a_1 = 1, \; a_n \in \mathbb{Z}.$$

These congruences are [2-9]:

$$(2.1) \qquad a_{np} - Tr(\pi_p)a_n + b_p \cdot a_{n/p} \equiv 0 \bmod p^s$$

if $n \equiv 0 \bmod p^{s-1}$, where $Tr(\pi_p)$ is a trace of Frobenius of E mod p; $b_p = 1$, if the reduction E mod p is good; and $b_p = 0$ if the reduction of E mod p is bad. Also it is well known (Manin [26] seems to be the first explicit source) that the values

$$P_{\frac{p-1}{2}}(\delta) \bmod p$$

of Legendre polynomials determine mod p the trace of the Frobenius of the curve E_δ for any δ mod p (in particular, for rational δ). Taking this and (2.1) into account, one derives from (2.1) congruences for values of Legendre polynomials $P_{\frac{p^k-1}{2}}(\delta)$ for δ mod p. On the other hand, one can consider the Jacobi quartic E_δ over the field $\mathbf{F}_p(\delta)$ (Honda [27], Yui [28]). One arrives at congruences for the Legendre polynomials themselves. The first of such congruences is known as Schur's congruence and had been studied for some time in connection with the outstanding problem of irreducibility of Legendre polynomials. Schur's congruence (an analog of a more widely known Lucas congruence) states that for an integer n having p-adic expansion $n = d_0 + d_1p+\ldots+d_kp^k$, one has a polynomial congruence [29]

$$P_n(x) \equiv P_{d_0}(x)\cdot P_{d_1}(x^p)\ldots P_{d_k}(x^{p^k}) \bmod p. \tag{2.2}$$

Using formal groups Honda [27] and Yui [28] derived new congruences mod p^k for $k > 1$ for special series of indices of Legendre polynomials.

New congruences for $P_{\frac{p^k-1}{2}}(\delta)$ are connected with Hazewinkel generators v_n of a p-typical formal group [2]. These congruences arise from the K-theoretic demands of Ravenel and Landweber-Stong see [30]. The Hazewinkel generators in this situation are associated with the logarithm $\ell_p(z) = \Sigma_{n=0}^{\infty} P_{\frac{p^n-1}{2}}(\delta)\cdot p^{-n}\cdot z^{p^n}$ and are defined as:

$$P_{\frac{p^n-1}{2}}(\delta) = \Sigma_{i=0}^{n-1} p^{n-1-i} v_{n-i}^{p^i}\cdot P_{\frac{p^i-1}{2}}(\delta),$$

so that one has:

$$v_1 = P_{\frac{p-1}{2}}(\delta).$$

The interpretation of new congruences lies in the identification of $v_n (\mathrm{mod}\ v_1)$.

To understand the geometric meaning of these congruence we recall that $\mathrm{Tr}(\pi_p) \equiv P_{\frac{p-1}{2}}(\delta)$ mod p. Thus congruences mod v_1, or rather mod(v_1,p) correspond to the study of elliptic curves E_δ that are supersingular at p, i.e. $\mathrm{Tr}(\pi_p) \equiv 0 (\mathrm{mod}\ p)$. We refer readers to [30] for the treatment of congruences between some Legendre polynomials mod v_1 from the point of view of geometry of supersingular elliptic curves. One of their crucial properties is the fact that elliptic curves supersingular at p are always defined over $\mathbb{F}_{p^2}$.

The first congruence that was needed in [30] is:

$$\frac{1}{p}\{P_{\frac{(p^2-1)}{2}}(\delta) - P_{\frac{(p-1)}{2}}(\delta)^{1+p}\} \equiv (-1)^{\frac{p-1}{2}} \mod(p, P_{\frac{p-1}{2}}(\delta)). \tag{2.3}$$

Note that in view of the Schur congruences (2.2), the left side is a p-integral polynomial. It turns out that there is a sequence of congruences similar to (2.3) for higher powers of p. In order to represent them in the compact form, we put for $k \geq 2$:

$$\mathrm{Rem}(p^k) \overset{\mathrm{def}}{=} P_{\frac{p^k-1}{2}}(\delta) \mod P_{\frac{p-1}{2}}(\delta) \tag{2.4}$$

to be the remainder of two polynomials in δ. Then we present the following congruences. First:

$$\mathrm{Rem}(p^2) \equiv (-1)^{\frac{p-1}{2}} p \mod p^3. \tag{2.5}$$

This is an improvement over (2.3) because of p^3 instead of p^2 in the right side of (2.5). Next we have a list of similar congruences for higher powers of p

$$\mathrm{Rem}(p^3) \equiv 0 \mod p^6; \tag{2.6}$$

$$\mathrm{Rem}(p^4) \equiv p^2 \mod p^3. \tag{2.7}$$

Next, in fact, <u>we conjecture</u> that the following is true.

$$\mathrm{Rem}(p^4) \equiv p^2 \mod p^4; \tag{2.7'}$$

$$\mathrm{Rem}(p^5) \equiv 0 \mod p^7; \tag{2.8}$$

$$\mathrm{Rem}(p^6) \equiv (-1)^{\frac{p-1}{2}} p^3 \mod p^4. \tag{2.9}$$

Conjecturally one has the following (best possible) congruences:

$$\begin{cases} \mathrm{Rem}(p^{2k+1}) \equiv 0 \mod p^{k+5} & \text{for } k \geq 1, \\ \mathrm{Rem}(p^{4k}) \equiv p^{2k} \mod p^{2k+2} & \text{for } k \geq 1, \\ \mathrm{Rem}(p^{4k+2}) \equiv (-1)^{\frac{p-1}{2}} p^{2k+1} \mod p^{2k+3} & \text{for } k \geq 0 \end{cases} \tag{2.10}$$

some of which we can prove.

§3. Abelian Varieties and Primality Testing

It is well known that one of the best (though, only necessary) tests of primality of p is based on Fermat's Little Theorem

$$x^p \equiv x \pmod p$$

for any $x \bmod p$. This test, say, for $x = 3$, is also an efficient "practically sufficient" test of primality of a prime suspect p, provided p has no small factors [31]. Though the inverse to the Little Fermat Theorem is incorrect (for Carmichael's numbers), there are various ways to obtain sufficient conditions of primality from the structure of the multiplicative group of $(\mathbb{Z}/n\mathbb{Z})^*$. E.g. one has the famous "n-1 primality criterion" (Pocklington, 1914; Selfridge, see [32]).

If for every prime divisor p of $n-1$ there exists $x_p \bmod n$ such that

$$x_p^{n-1} \equiv 1 \bmod n,$$

but

$$x_p^{(n-1)/p} \not\equiv \bmod n,$$

then n is prime. Moreover, it is apparently sufficient to check only a factor of $n - 1 = F_1 \cdot R_1$, i.e. to look at $p | F_1$ when $F_1 > \sqrt{n}$.

It was noticed some time ago (Lucas-Lehmer,...) that one can derive a similar criterion based on factors of $n + 1 (n^2+1,\ldots)$, if one considers the so called Lucas sequences.

To see how these and other criteria of primality are derived one can reformulate the Little Fermat Theorem in terms of a Frobenius acting on an algebraic (formal) group over $\mathbb{F}_q$ [33]:

<u>Little Fermat Theorem:</u> Let an algebraic group A be defined over $\mathbb{F}_q$ (e.g. A is a reduction of an algebraic group scheme defined over $\mathbb{Z}[1/N]$ for $(N,p) = 1$), and let $P(x)$ be a characteristic polynomial of a Frobenius automorphism F_q of A. Then for an arbitrary X of A, the sequence of multiples

$$X_m = [m]_A(X) : m \in \mathbb{Z}$$

in the group law of A has a rank of apparition of p, and this rank, $\tau(p)$, divides $P(1)$. This means that

$$X_m = \bar{0} \text{ in } A/\mathbb{F}_q$$

if and only if

$$\tau(p) | m$$

for $\tau(p) | P(1)$.

In these notations the Lucas sequences correspond to the multiplicative formal group law

$$F_{\sqrt{D}}(x,y) = x + y + \sqrt{D}\cdot x\cdot y.$$

Similar to $n \pm 1$ criteria of primality we have presented [33] sufficient (and necessary) criteria of primality based on the divisibility of orders of elliptic curves (and CM-Abelian varieties) A defined over $\mathbb{Z}/N\mathbb{Z}$. E.g. let $N = \prod_{i=1}^{k} p_i^{\alpha_i}$, where p_i are (distinct) divisors of n (> 3), and let E be an elliptic curve defined over $\mathbb{Z}/N\mathbb{Z}$ (over $\mathbb{F}_{p_i}$ for all p_i). If one denotes by $\varphi_E(\cdot)$ the analog of Euler function: $\varphi_E(N) = \prod_{i=1}^{k} p_i^{\alpha_i - 1}(p_i + 1 - a_{p_i}(E))$, where $a_{p_i}(E)$ are traces of Frobenius of E mod p_i, then $\varphi_E(N)$ coincides with the order of $E(\text{mod } N)$ if and only if N is prime.

Goldwasser et al. suggested a (probabilistic) primality test, based on elliptic curves, where to prove the primality of a suspect prime n, one had to find (by random search) an elliptic curve mod n, whose order is $2q$, where q is a (suspect) prime. Then one checks the primality of q, etc. The polynomiality of the running time of this algorithm depends on conjectures of distribution of primes. Moreover, this algorithm, and similar ones [34], are hard to implement for large n, because they depend on fast methods of computation of numbers of points (mod $n = p$) on elliptic curves over $\mathbb{F}_p$. Though there is Schoof's [35] algorithm that allows to compute this number in $O(\log^6 p)$ steps, its realization is unrealistic for p having a hundred digits. It is much more efficient to start with elliptic curves and Abelian varieties with complex multiplication, where one has explicit formulas for the traces of Frobenius. Atkin constructed an algorithm that uses modular equations (when one can have explicit formulas) that can work with primality proving for 200-400 digit long numbers see [34]. The best results though are achieved using CM varieties see [33]. Also to bound complexities, instead of modular curves that have to be reduced mod p, but are to be determined in complex number arithmetic, one can work with Drinfeld curves corresponding to elliptic moduli over finite fields.

§4. Algorithms of polynomial multiplication and algebraic theory of linear codes.

Algebraic complexities were introduced to describe the number of operations: multiplications (multiplicative complexity) and additions (additive complexity) in familiar algebraic structures: rings of matrices and rings of polynomials. While the operation of addition in such rings is straightforward, it is by no means obvious how to realize the operation of multiplication in these rings in the least number of operations on scalars, particularly in the least number of multiplications.

The algebraic complexity problem which is the richest in its underlying structure is the problem of fast polynomial multiplication. Among other problems reducible to that one can mention: fast multiplications of multiple-precision numbers, gcd's in polynomial rings, Hankel matrix multiplication, computation of rational and Padé approximations, computation of finite Fourier transformations,...etc. Significant progress in this problem, due to Winograd, Fiduccia, Strassen and others, was concentrated mainly on minimal multiplicative complexities of polynomial multiplication over <u>fields</u>. In practical implementation of fast multiplication of multiple-precision integers algebraical complexities have to be considered over a ring. This ring is $\mathbb{Z}$ or $\mathbb{Z}[1/2]$.

In this chapter we show how low complexity algorithms of polynomial multiplication are connected with interpolation methods (Toom-Cook-Winograd algorithms for infinite fields), and we present new ideas of interpolation on algebraic curves of positive genus, that are applied to new low multiplicative complexity polynomial multiplication algorithms (particularly over finite fields).

Let us start with definitions of multiplicative complexity of a system of bilinear forms [36,37], [31].

Let $\mathbb{A}$ be a ring (of scalars), and suppose given s bilinear forms in variables $\bar{x} = (x_1,\ldots,x_m)$ and $\bar{y} = (y_1,\ldots,y_n)$ with coefficients from $\mathbb{A}$:

$$(4.1)\qquad z_k = \sum_{i=1}^{m} \sum_{j=1}^{n} t_{i,j,k}\, x_i y_j: \ k = 1,\ldots,s.$$

One of the most widely used definitions of the multiplicative complexity of computation of a system (4.1) over $\mathbb{A}$ is that of the rank of the $m \times n \times s$ tensor $(t_{i,j,k})$ over $\mathbb{A}$ (Strassen). The definition of a rank of a tensor $T = (t_{i,j,k})$ over $\mathbb{A}$ proceeds as follows [31]. A nonzero tensor T is said to be of rank 1 over $\mathbb{A}$, if there are three $\mathbb{A}$-vectors $(a_1,\ldots,a_m),(b_1,\ldots,b_n),(c_1,\ldots,c_s)$ such that

$$t_{i,j,k} = a_i \cdot b_j \cdot c_k \quad \text{for all} \quad i,j,k.$$

Then, the rank of a tensor T over $\mathbb{A}$ is the minimal number, $\delta_{\mathbb{A}}$, such that T is expressable as a sum of $\delta_{\mathbb{A}}$ rank 1 tensors over $\mathbb{A}$.

This scheme of evaluation of (4.1) is called a bilinear scheme (normal or noncommutative). In this scheme for the evaluation of a system of bilinear forms (4.1) one forms δ products of linear combinations of x's and y's:

$$(4.2a)\qquad w_\ell = (a_{\ell 1}x_1 + \ldots + a_{\ell m}x_m)(b_{\ell 1}y_1 + \ldots + b_{\ell n}y_n),$$

$\ell = 1,\ldots,\delta$; and then obtains z's in (4.1) as linear combinations of these products:

$$(4.2b)\qquad z_k = c_{k1}w_1 + \ldots + c_{k\delta}w_\delta: \ k = 1,\ldots,s.$$

In this definition of multiplicative complexity one counts only <u>non-scalar</u> (essential) multiplications, i.e. δ multiplications in (4.2a), but does not count scalar multiplications by $a_{\ell i}, b_{\ell j}, c_{k\ell}$-elements of $\mathbb{A}$.

Introducing "layers" $T_k = (t_{i,j,k})_{i=1}^{m}\,{}_{j=1}^{n}$ of the tensor $T = (t_{i,j,k})$, one can reformulate the definition of rank δ of T as the minimal number δ such that all $m \times n$ matrices T_k lie in a linear span over $\mathbb{A}$ of δ rank one matrices (of the form $\bar{a}_\ell \otimes \bar{b}_\ell$: $\ell = 1,\ldots,\delta$--called "dyads").

For future reference, we express the bilinear algorithm (4.2a)-(4.2b) of computation of bilinear forms (4.1) in the following algebraic form

$$\bar{z} = C\cdot(A\bar{x} \otimes B\bar{y}) \tag{4.3}$$

for matrices $C = (c_{k\ell})(\in M_{s\times\delta}(\mathbb{A}))$, $A = (a_{\ell i})\ (\in M_{\delta\times m}(\mathbb{A}))$, $B = (b_{\ell j})\ (\in M_{\delta\times n}(\mathbb{A}))$.

If one extends the ring of scalars by inverting primes and by adding algebraic numbers (most notable roots of unity), one can significantly reduce the multiplicative complexity of computation of systems of bilinear forms. This is particularly visible for polynomial multiplications. Unfortunately, simultaneously the total computational complexity of the algorithms (as reflected in the total number of bit operations) becomes unbearably high. That is why the "least complex" algorithms of polynomial multiplication by Toom-Cook-Winograd [31,38] are unattractive in practice for relatively large degrees. We describe briefly these algorithms because they provide clues to the use of interpolation technique.

The basic idea underlying the Toom-Cook method is to use interpolation is reconstruction of coefficients of a product of two polynomials. Namely, let us consider the bilinear problem (4.1) corresponding to the multiplication of two polynomials of degree m - 1 and n - 1, respectively:

$$P(t) = \sum_{i=1}^{m} x_i t^{i-1}, \qquad Q(t) = \sum_{j=1}^{n} y_j t^{j-1}. \tag{4.4}$$

We are interested in the efficient determination of coefficients z_k of a polynomial product:

$$R(t) \overset{\text{def}}{=} P(t)\cdot Q(t), \quad R(t) = \sum_{k=1}^{m+n-1} z_k t^{k-1}. \tag{4.5}$$

An obvious scheme of evaluation of (4.5) takes mn multiplications (but no scalar multiplication). Toom noticed, however, that one can identify R(t), if one knows its values at m + n - 1 distinct points α_k: k = 1,...,m + n - 1 in terms of the Lagrange interpolation formula:

$$R(t) = \sum_{k=1}^{m+n-1} R(\alpha_k)\cdot \frac{\prod_{\ell\neq k}(t-\alpha_\ell)}{\prod_{\ell\neq k}(\alpha_k-\alpha_\ell)}. \tag{4.6}$$

Thus if one assumes the ring $\mathbb{A}$ to <u>include</u> α_i and $(\alpha_i-\alpha_j)^{-1}$ (for $i\neq j$), then one can reconstruct all $m+n-1$ coefficients z_k of $R(t)$ from $R(\alpha_k)$ using scalar multiplications only. To compute $R(\alpha_k)$ one needs only $m+n-1$ nonscalar multiplications:

$$R(\alpha_k) = P(\alpha_k)\cdot Q(\alpha_k), \tag{4.7}$$

where $P(\alpha_k)$ and $Q(\alpha_k)$ are computed by the Horner scheme in (4.4) (or by its fast version) in scalar(from $\mathbb{A}$) multiplications only.

The Toom-Cook scheme was generalized by Winograd[39-39], who showed that one can evaluate $R(t)$ mod arbitrary relatively prime polynomials $D_\alpha(t)$, and then bring the results together using the Chinese remainder theorem and without any nonscalar multiplications to obtain $R(t) \bmod \prod_\alpha D_\alpha(t)$. Toom-Cook took $D_\alpha(t) = t - \alpha$, while Winograd considered $D_\alpha(t)$ as cyclotcmic polynomials. Optimal results of polynomial multiplication and multiplications in algebras $\mathbb{A}[x]/(T(x))$ (i.e. polynomial multiplication mod$T(x)$), when $\mathbb{A} = K$ is an <u>infinite</u> <u>field</u>, are described by the following Winograd's

<u>Theorem 4.1</u>: Let K be an infinite field. Then the multiplicative complexity, $\delta_K(m,n)$, of multiplication of polynomials of degrees $m-1$ and $n-1$ over $K[x]$ is $m+n-1$ exactly. The multiplicative complexity, $\delta_K(T)$, of multiplication of polynomials in $K[x]$ mod $T(x)$ for $T(x) \in K[x]$ is equal to $2n-k$, where k is a number of distinct irreducible factors of $T(x)$ in $K[x]$.

Though these results are astonishingly strong, none of them can be applied in practice (starting from $m \geq 4$, $n \geq 4$) because in any of these schemes realizing the minimal multiplicative complexity over, say, $\mathbb{Q}$, the scalar multiplications--not counted as "actual multiplications"—involve scalars that are too large to be counted as just a few more additions. Also one has to divide by relatively large integers having no advantageous binary structure.

For example, in the notations of (4.3), let us describe the multiplication routine over $\mathbb{Q}$, which realizes the minimal number of multiplications of a polynomial of degree two by a polynomial of degree three (see [31]) $(x_1+x_2t+x_3t^2)(y_1+y_2t+y_3t^2+y_4t^3)$:

$$A = \begin{pmatrix} 1 & 1 & 1 & 1 & 1 & 1 \\ 0 & 1 & 2 & 3 & 4 & 5 \\ 0 & 1 & 4 & 9 & 16 & 25 \end{pmatrix}, \quad B = \begin{pmatrix} 1 & 1 & 1 & 1 & 1 & 1 \\ 0 & 1 & 2 & 3 & 4 & 5 \\ 0 & 1 & 4 & 9 & 16 & 25 \\ 0 & 1 & 8 & 27 & 64 & 125 \end{pmatrix},$$

$$C = \begin{pmatrix} 120 & 0 & 0 & 0 & 0 & 0 \\ -274 & 600 & -600 & 400 & -150 & 24 \\ 225 & -770 & 1070 & -780 & 305 & -50 \\ -85 & 355 & -590 & 490 & -205 & 35 \\ 15 & -70 & 130 & -120 & 55 & -10 \\ -1 & 5 & -10 & 10 & -5 & 1 \end{pmatrix} \cdot \frac{1}{120}$$

To get an idea of the size of scalars involved in the multiplication schemes of minimal complexity, we give the following lower bound on the maximum of sizes of elements of A, B and C: $\max(|a_{\ell i}|,|b_{\ell j}|,|c_{k\ell}|) \geq \max((m+n-3)!,(m+n-2)^{\max(n,m)-1})$ in the multiplication of polynomials of degrees n - 1 and m - 1 with the field of scalars $\mathbb{A} = \mathbb{Q}$. Also if $A \in M_{(n+m-1)\times m}(\mathbb{Z})$, $B \in M_{(n+m-1)\times n}(\mathbb{Z})$, $C = \frac{1}{d}\cdot C_1$, $C_1 \in M_{(n+m-1)\times(n+m-1)}(\mathbb{Z})$, then $\max(|a_{\ell i}|) \geq (m+n-3)^m$, $\max(|b_{\ell j}|) \geq (m+n-3)^n$, $\max(|c_{k\ell}|) \geq (m+n-2)!$, $d \geq (m+n-3)!$

Not only the large sizes of scalars make the number of additions prohibitively--exponentially--large, but the coefficients becomes large integers requiring fast bignum multiplications. The total number of operations, counted in terms of single precision additions and multiplications, in minimal multiplicative complexity algorithms of Theorem 4.1 significantly exceeds the number of operations in the standard high school methods of polynomial multiplication. This makes the minimal multiplication complexity algorithms of Theorem 4.1 unsuitable for practical implementation. The minimal multiplication routines of Theorem 1 found their place, though, in Winograd's [31,38] short prime length DFT algorithms; they are well suited for iterations when one uses them only for short data sizes.

It is more efficient in practice to use fast multiplication routines with as little division by scalars as possible. From the point of view of hardware realization it is preferable to have division by power of 2 only, i.e. one should consider minimal multiplication schemes over $\mathbf{A} = \mathbf{Z}[1/2]$. It was realized some time ago (Schonhage, Strassen [40], Winograd [38-39] Nussbaumer [41]) that one can achieve fast multiplication of polynomials, with division by 2 only, if one considers polynomial multiplications modulo cyclotomic polynomials $(x^{2^n}-1)$. In this approach: $x^{2^n}-1 = \prod_{j=0}^{n-1}(x^{2^j}+1)\cdot(x-1)$ and one brings together via the Chinese remainder theorem the remainders of polynomial multiplications modulo cyclotomic factors $(x^{2^j}+1)$: $j = 0,\ldots,n-1$. This general scheme, coupled with fast Fourier transforms in finite fields (modulo factors of Fermat numbers) was used first by Schonhage-Strassen [40] to achieve asymptotically fast multiplication of large integers. In that method one can multiply two n-bit integers in time $O(n \log n \log\log n)$ (i.e. in that many bit operations). Variations of the same method constitute a fast convolution algorithm (Nussbaumer [41]) according to which two polynomials can be multiplied modulo x^N-1 in at most $3\,N \log_2 N$ nonscalar multiplications (with $N = 2^{2^n}$) over the ring $\mathbf{Z}[1/2]$ of scalars.

We see that linear upper bounds for multiplicative complexities of polynomial multiplication (given, say, in Theorem 4.1) imply subexponential lower bounds for additive complexities of these algorithms. The best upper bounds for total (both multiplicative and additive) algebraic complexities of computation of polynomial multiplication via certain versions of FFT are $O(n\log n)$ for polynomials of degrees bounded by n, even though divisions by powers of 2 are necessary in this scheme. No nonlinear lower bound is known for algebraic complexities in this problem.

To see what are $\mathbf{Z}$-algorithms of fast polynomial multiplication, one has to consider first the reductions mod p, and to look at algebraic complexities of polynomial multiplications over $\mathbf{A} = \mathbf{F}_p$, particularly over $\mathbf{A} = \mathbf{F}_2$.

Over finite fields there is no simple answer to the minimal multiplicative complexity of polynomial multiplication like one given in Theorem 4.1 for infinite fields. If $\delta_{\mathbf{A}}(m,n)$ denotes the minimal multiplicative complexity of multiplication of polynomials of degrees $m-1$ and $n-1$, respectively, with a ring of constants $\mathbf{A}$, then we always have $\delta_K(m,n) \geq m+n-1$ for an arbitrary field K of scalars, but this inequality becomes equality <u>only</u> when K has at least $m+n-2$

elements.

At it turns out, lower bounds for multiplicative complexities over finite fields, and, as a consequence, over the ring $\mathbb{Z}$ of scalars, are much stronger than the one given above. To obtain them, though,the algebraic theory of linear codes has to be introduced.

Let us recall the basics of linear codes [42]. In the theory of linear codes one considers vector spaces A^n of dimension n over a finite field $\mathbb{F}_q$, where "an alphabet" A consists of elements of the field $\mathbb{F}_q$. A linear subspace of A^n is called a linear code. A linear code is the null space of a parity check matrix of the code, and a basis of the code form the rows of the matrix, called a generator matrix. In addition to n, two more parameters: k and d are associated with a code (called an $[n,k,d]$-code). First, we denote by k the dimension of the code over $\mathbb{F}_q$. Second, by the weight of the code, denoted d, we understand the minimal number of nonzero coordinates of all nonzero vectors from the code with respect to a fixed basis of the space A^n. Hamming's problem consists of the construction of codes having the largest possible weights d for given n and k. In the context of this problem we define $N(k,d)$ as the least integer n, such that there exists a code with given n and k, of weight d.

For a long time the best upper bounds on $N(k,d)$ followed from the Gilbert-Varshamov bound that establishes (but does provide for effective construction) the existence of good codes (i.e. d is large for given n and k). In this and similar bounds one puts $H_q(x) \doteq -x \log_q x - (1-x)\log_q(1-x)$. Then the Gilbert-Varshamov bound [42] proves the existence of $[n,k,d]$-linear codes over $\mathbb{F}_q$ such that for $R \overset{\text{def}}{=} k/n$, $\delta \overset{\text{def}}{=} d/n$ one has $R \geq 1 - \delta \log_q(q-1) - H_q(\delta)$.

V. Goppa in the 70's and 80's constructed codes that meet (and sometimes exceed) the Gilbert-Varshamov bound using algebraic curves over finite fields. Goppa's first construction (described in [42]) involved only rational curves over $\mathbb{F}_q$. In this construction one chooses n distinct elements of $\mathbb{F}_q$: $x_1,\dots,x_n$ and considers the following linear subspace (code) in $\mathbb{F}_q^n$:
$C = \{(f(x_1),\dots,f(x_n)) \in \mathbb{F}_q^n : f \in \mathbb{F}_q[t], \deg(f) \leq k-1\}$. This code has dimension k, where $k \leq n \leq q$, and its weight is $d = n + 1 - k$.

Later Goppa realized that one can look at linear systems on

arbitrary (nonsingular) curves, see his review [43]. Identically to the case of rational curves above, let us consider a (nonsingular) curve Γ over $\mathbb{F}_q$ and two divisors D and G on Γ, both positive and with disjoint supports: $D = \Sigma_{i=1}^{n} P_i$, where P_i are distinct points from $\Gamma(\mathbb{F}_q)$. Then the linear code associated with $(\Gamma;D,G)$ is defined as the image of the natural map $f \in \mathcal{L}(G): \to (f(P_1),\ldots,f(P_n)) \in \mathbb{F}_q^n$. According to Riemann-Roch, whenever $\deg(G) > 2g - 2$, the dimension k of this code is $\deg(G) - g + 1$. The weight d of this code is $\geq n - \deg(G)$. Goppa showed that in order to construct optimal codes in this approach, i.e. with large d for given n, k (as $n \to \infty$) one needs families of curves over $\mathbb{F}_{q^m}$ with maximal "allowable" number of points for a fixed genus $g (g \to \infty)$. Goppa himself found several interesting classes of examples of such curves coming from Fermat curves $X^m + Y^m + Z^m = 0$. We return to this problem in the next section, when we use Goppa-like codes for the construction of new low complexity polynomial multiplication algorithms.

We will demonstrate the relationship between (multiplicative) complexity of multiplication in k-algebras and the Hamming problem for linear codes over k in the most general situation. For this we represent bilinear algorithms of computation of multiplication in k-algebras in the coordinate-free form [37].

Let $\mathbb{A}$ be a k-algebra of dimension n over k (an arbitrary field) with the basis $e_1,\ldots,e_n$. If we have a multiplication table $e_i e_j = \Sigma_{m=1}^{n} C_{ij}^{m} e_m$ for $C_{ij}^{m} \in k$ $(i,j = 1,\ldots,n)$, then the multiplication rule in $\mathbb{A}$ can be written in bilinear form. For two elements of $\mathbb{A}$: $\Sigma_{i=1}^{n} x_i e_i$ and $\Sigma_{j=1}^{n} y_j e_j$ we have $(\Sigma_{i=1}^{n} x_i e_i)\cdot(\Sigma_{j=1}^{n} y_j e_j)$ $= \Sigma_{m=1}^{n} z_m e_m$ with $z_m = \Sigma_{i,j} x_i y_j C_{ij}^{m}$ as in (4.1). In the coordinate-free form we associate with the multiplication in $\mathbb{A}$ over k: $X: \mathbb{A} \times \mathbb{A} \to \mathbb{A}$ a 3-tensor $t_{\mathbb{A}} \in \mathbb{A}^* \otimes \mathbb{A}^* \otimes \mathbb{A}$. The rank of this tensor is the minimal number δ (the multiplicative complexity $\delta = \delta_k(\mathbb{A})$) such that $t_{\mathbb{A}}$ is represented as a sum of δ rank one tensors

$$t_{\mathbb{A}} = \Sigma_{\ell=1}^{\delta} u_\ell \otimes v_\ell \otimes w_\ell \tag{4.7}$$

for $u_\ell \otimes v_\ell \otimes w_\ell$ in $\mathbb{A}^* \otimes \mathbb{A}^* \otimes \mathbb{A}$. One can define layers of $t = t_{\mathbb{A}}$ as $t_x = \Sigma_{\ell=1}^{\mu} u_\ell(x) v_\ell \otimes w_\ell$ and $t^y = \Sigma_{\ell+1}^{\delta} v_\ell(y) u_\ell \otimes w_\ell$ for $x \in \mathbb{A}$, $y \in \mathbb{A}$: t_x and t_y are linear mappings $\mathbb{A} \to \mathbb{A}$. According to the definition of the multiplication tensor t,

$$t_x = L_x, \quad t^y = R_y \tag{4.8}$$

where L_x and R_y are left and right multiplications by x and y respectively in $\mathbb{A}$. We refer to [44] for the relationship between the layers of t with respect to the third index and the left multiplication in $\mathbb{A}$.

Corollary 4.2: If $\mathbb{A}$ is a $\mathbb{F}_q$-algebra of dimension n over $\mathbb{F}_q$ and without zero divisors, then every realization of multiplication in it over $\mathbb{F}_q$ as a bilinear algorithm with $\delta = \delta_{\mathbb{F}_q}(\mathbb{A})$ nonscalar multiplications over $\mathbb{F}_q$ gives rise to a $[\delta,n,n]$-linear code over $\mathbb{F}_q$.

Proof: Let us consider a bilinear algorithm (4.7) over $k = \mathbb{F}_q$. We define as a linear code C the set of all vectors $\bar{u}(x) = (u_\ell(x):$ $\ell = 1,\dots,\mu)$ in k^δ. The layer t_x of $\mathbb{A}$ with respect to a nonzero $x \in \mathbb{A}$ is $L_x = \Sigma_{\ell=1}^{\delta} u_\ell(x) v_\ell \otimes w_\ell$. Because x is not a zero divisor, $L_x \in G\ell(\mathbb{A})$. In particular, it means that there are at least n nonzero $u_\ell(x)$ for $\ell = 1,\dots,\delta$ for every nonzero $x \in \mathbb{A}$. This is equivalent to the statement that C is a $[\delta,n,n]$-code.

Corollary 4.2 includes, in particular, all finite extensions of prime fields.

The proof of Corollary 4.2 also provides important clues as to matrices A, B and C in the algebraic form (4.3) of the algorithm of multiplication in k-algebra $\mathbb{A}$.

Using Corollary 4.2 and known lower bounds on $N(k,d)$ from the theory of linear codes, one can bound from below the multiplicative complexity of polynomial multiplication. One of the best linear code bounds is (MRRW) proved in [45]. According to this bound for $q = 2$, if $n \to \infty$ and there is a sequence of $[n,k,d]$-codes with $d/n \to \delta$, then $R \overset{\text{def}}{=} k/n \le H_2(1/2 - \sqrt{\delta-\delta^2})$. This bound was used in [46] to bound the multiplicative complexity $\delta_{\mathbb{F}_2}(n,m)$ of multiplication over $\mathbb{F}_2$ of polynomials of degrees $n-1$ and $m-1$ respectively: $\delta_{\mathbb{F}_2}(n,n) \ge 3.52\cdot n$, and $\delta_{\mathbb{Z}}(n,n) \ge 3.52\cdot n$ for a sufficiently large n.

Our results show that similar lower bounds hold for multiplicative complexity of multiplication in finite extensions of $\mathbb{F}_q$. Indeed, let $\mathcal{K} = \mathbb{F}_q[t]/p(t)$, for an irreducible polynomial $p(t)$ of degree n in $\mathbb{F}_q[t]$, so $\mathcal{K} \cong F_{q^n}$; and let $\delta_{\mathbb{F}_q}(\mathcal{K})$ denote the minimal multiplicative complexity in the field $\mathcal{K}$ over $\mathbb{F}_q$. Then Corollary 4.2 implies that there exists a $[\delta_{\mathbb{F}_q}(\mathcal{K}), n, n]$-linear code over $\mathbb{F}_q$. Combining this with the bound (MRRW) we deduce the lower bound

$$\delta_{\mathbb{F}_2}(\mathcal{K}) \geq 3.52\ n \tag{4.9}$$

for sufficiently large $n = [\mathcal{K}:\mathbb{F}_2]$.

For $q > 2$ one gets less sharp bounds from known bounds on optimal codes. For example, the Plotkin bound [42] implies

$$\delta_{\mathbb{F}_q}(\mathcal{K}) \geq (2 + \frac{1}{q-1}) \cdot n$$

as $n = [\mathcal{K}:\mathbb{F}_q] \to \infty$.

We found an interesting phenomenon for polynomial multiplication over finite fields drastically different from the infinite field case. In the case of an infinite field of constants k, Winograd's Theorem 4.1 shows that multiplication mod an irreducible polynomial $p(t)$ of degree n takes as many essential multiplications as that of multiplication of two polynomials of degree $n-1$. It is no longer true in the finite field case, when we always have $\delta_k(k[t]/(p(t))) + C \cdot n < \delta_k(n,n)$ for a positive constant $C(= C(k))$ and an arbitrary polynomial $p(t)$ of degree n over a finite field k. We conjecture, that in fact

$$\lim_{\substack{n\to\infty \\ [\mathcal{K}:k]=n}} \frac{\delta_k(n,n)}{\delta_k(\mathcal{K})} = 2.0.$$

The lower bounds for multiplicative complexities of polynomial multiplication are always linear in n, and seem far from the best nonlinear upper bound $O(n\log n)$ for finite field polynomial multiplication. In the next chapter we show that the upper bound can be brought down to a linear one with the constant comparable to that of the lower bound.

§5. Interpolation on algebraic curves and its application to polynomial multiplication algorithms.

How can one represent a rational function (for example, a

polynomial)? It can be represented as $P(t)/Q(t)$, in terms of its residues at singularities--as $c_0 + \Sigma_{i=1}^{n} c_i/(t-\alpha_i)$, or by its values at a given set of points. The last representation is the interpolation. Interpolation formulas are the basis of the Toom-Cook algorithm of the fast polynomial multiplication. Apparently the interpolation algorithm always has the lowest multiplicative complexity. The most important polynomial multiplication problem is that of multiplication in a finite extension $\mathcal{K} = k[t]/((p(t)))$ for a fixed $p(t) \in k[t]$ and a field of constants k. Let $\delta_k(\mathcal{K})$ be the multiplicative complexity of multiplication in $\mathcal{K}$ over k. For an infinite k and $\mathcal{K}$ of degree n over k, $\delta_k(\mathcal{K}) = 2n - 1$ by Theorem 4.1, and all algorithms realizing this multiplicative complexity are interpolation algorithms, interpolating products $x(t) \cdot y(t) \bmod p(t)$ at $2n-1$ distinct points of $k\mathbb{P}^1$. What happens if k is finite, say $\mathbb{F}_2$? There are not enough points to interpolate at. What usually is done is an extension of the field of constants (till there are enough points for interpolation) and the telescopic reduction of multiplication in the composite extension of fields. The key here is a trivial "multiplication rules":

$$\mu_k(\mathcal{K}) \le \mu_k(\mathcal{L}) \cdot \mu_{\mathcal{L}}(\mathcal{K}) \tag{5.1}$$

where $k \subset \mathcal{L} \subset \mathcal{K}$.

We suggest a novel way of interpolation by means of representing the set of points where one interpolates as a divisor on an algebraic curve. To represent this interpolation in a more abstract way one has to use the language of places from the theory of algebraic function of one variable, see [47].

By a place of a field K we understand an isomorphism $\varphi: K \to \Sigma \cup \{\infty\}$, where Σ is a field and $\varphi(a) = \infty$, $\varphi(b) \ne 0,\infty$ for some $a,b \in K$. There is an obvious correspondence between places and valuations (which we will use). All places on an algebraic function field are extensions of places from corresponding rational function fields, and places of a rational function field $K = k(x)$ over a field k correspond either to (i) irreducible polynomials $p(x)$ in $k[x]$, or to (ii) x^{-1}. We deal only with algebraic function fields in one variable. Any such field K over the field of constants k can be represented in the form $K = k(x,y)$, where x is (any) transcendental element of

K over k, and $1,\dots,y^{d-1}$ is the basis of K over $k(x)$, $[K: k(x)] = d$. For an arbitrary place $\wp$ of K let $k_\wp$ be a field such that $\wp$ is an isomorphism onto $k_\wp \cup \{\infty\}$. We denote by $v_\wp$ the normed valuation with values in $\mathbb{Z}$, corresponding to $\wp$. The degree $f_\wp$ of $k_\wp$ over k is called the degree of $\wp$. A divisor of K is an element of the free Abelian group generated by the set of places of K. The places themselves are called prime divisors. We write divisors additively: $\mathfrak{a} = \Sigma_\wp v_\wp(\mathfrak{a})\cdot\wp$, where $v_\wp(\mathfrak{a})$ are integers among which only finitely many are nonzero. A divisor $\mathfrak{a}$ is called an integral one, if $v_\wp(\mathfrak{a}) \geq 0$ for every $\wp$. A divisor $\mathfrak{a}$ divides $\mathfrak{b}$, if $\mathfrak{b} - \mathfrak{a}$ is integral. The degree $d(\mathfrak{a})$ of a divisor $\mathfrak{a}$ is an integer $d(\mathfrak{a}) = \Sigma_\wp f_\wp v_\wp(\mathfrak{a})$. With every element $X \in K$ one associates a principal division $(X) = \Sigma_\wp v_\wp(x)\cdot\wp$.

The most important object is the vector space $L(-\mathfrak{a})$ over k consisting of all functions $X \in K$ such that the divisor $(X) + \mathfrak{a}$ is positive. If g denotes the genus of K see [47], then the dimension of $L(-\mathfrak{a})$ over k is determined by the Riemann-Roch theorem [47]. To formulate it, we denote by C an equivalence class of divisors in K (modulo the principal ones--(X) for $X \in K$), and by $N(C)$ the dimension of C, i.e. the maximal number of linearly independent integral divisors in this class. Then $N(C)$ is equal to $\dim L(-\mathfrak{a})$ for any $\mathfrak{a}$ from C. The Riemann-Roch theorem states that

(5.2) $$N(C) = d(C) - g + 1 + i(C),$$

where $i(C) = 0$ is the index of speciality of C, $i(C) = N(W-C)$, where W is the class of all differentials on K (canonical class, [47]). In particular, $N(C) = d(C) - g + 1$ if $d(C) > 2g - 2$ or $d(C) = 2g - 2$ and $C \neq W$.

In our applications, k is often a finite field of characteristic $p > 0$ with $q = p^m$ elements. If $\wp$ is any prime divisor of K--an algebraic function field with the field of constants k--then the number of elements in the residue field $k_\wp$ is called the norm of $\wp$ and is denoted as: $N(\wp) = q^{d(\wp)}$. This definition is extended to all divisors: $N(\mathfrak{a}) = q^{d(\mathfrak{a})}$. With an algebraic function field K/k one can associate a ζ-function:

(5.3) $$\zeta(K;s) = \Sigma_{\mathfrak{a}}(N(\mathfrak{a}))^{-s}, \ s > 1$$

where $\mathfrak{a}$ runs over all integral divisors of K/k. Let k have q elements. Then we have the Euler product

(5.4) $$\zeta(K;s) = \prod_{\mathcal{P}}(1 - N\mathcal{P}^{-s})^{-1},$$

where $\mathcal{P}$ runs over all prime divisors of K/k.

The Weil theorem (the Riemann hypothesis over finite fields) allows to express $Z(u) = \zeta(K; -\log u/\log q)$ in terms of eigenvalues of the Frobenius operator on K/k. Namely, there exists a polynomial $P_{2g}(u)$ with rational integer coefficients, $P_{2g}(u) = \prod_{i=1}^{2g}(1 - u\omega_i)$, such that

(5.5) $$Z(u) = P_{2g}(u)/(1-u)(1-qu).$$

For $i = 1,\ldots,2g$, $|\omega_i| = \sqrt{q}$ and q/ω_i is also one of ω_j's [48].

The interpolation idea on a curve corresponding to K/k proceeds as follows. To describe the elements of a finite extension $k_{\mathcal{P}}$ of k of degree $d(\mathcal{P})$ one looks for a positive divisor $\mathfrak{B}$ such that the natural mapping

$$\mathcal{P}: L(-\mathfrak{B}) \to k_{\mathcal{P}}$$

is onto. This way we find a basis of $k_{\mathcal{P}}$ over k among elements of $L(-\mathfrak{B})$. E.g. if $K = k(t)$ is a rational function field, and $\mathcal{P}$ corresponds to $p(t)$, one takes $\mathfrak{B} = (n-1)\cdot\infty$ for $n = \deg p(t)$, i.e. looks for a power basis $0,1,\ldots,t^{n-1}$ of an algebraic extension $k_{\mathcal{P}} = k[t]/(p(t))$ of k.

The multiplication law in the field $k_{\mathcal{P}}$ can be then represented in terms of the multiplication rule $L(-\mathfrak{B}) \times L(-\mathfrak{B}) \hookrightarrow L(-2\cdot\mathfrak{B})$. It remains now to reconstruct functions from $L(-2\mathfrak{B})$ be interpolation. For this one could take a set $\mathfrak{D}$ of divisors of the first degree (places of the first degree) such that $\mathrm{Card}(\mathfrak{D}) > 2d(\mathfrak{B})$. Any function from $L(-2\cdot\mathfrak{B})$ can be uniquely reconstructed by its values at places from $\mathfrak{D}$.

This general method is summarized in the following statements, where for an arbitrary integral (positive) divisor $\mathfrak{a}$ on K, we put $K/\mathfrak{a} \overset{\text{def}}{=} K \bmod \mathfrak{a}$ (and $X \equiv Y \bmod \mathfrak{a}$, if $(X-Y)$ is divisible by $\mathfrak{a}$).

Proposition 5.1: Let k be a field of constants, and let K be a function field over k of genus g with a prime divisor $\mathfrak{a}$ of degree n on K. Let $\mathfrak{B}$ be an integral divisor on K such that the natural mapping $j: L(-\mathfrak{B}) \to K/\mathfrak{a} = k_{\mathfrak{a}}$ is surjective. If $\mathfrak{D}$ is a set of prime divisors of the first degree on K and $\mathrm{Card}(\mathfrak{D}) > 2d(\mathfrak{B})$, then there exists a bilinear algorithm for multiplication in the field $k_{\mathfrak{a}} = K/\mathfrak{a}$ of degree n over k with the field of scalars k, whose multiplicative complexity is $\leq \dim_k L(-2\mathfrak{B}) \leq \mathrm{Card}(\mathfrak{D})$.

Proof: Since $j: L(-\mathfrak{B}) \to k_{\mathfrak{a}}$ is surjective, we can always choose a basis $f_1,\ldots,f_n$ of $K/\mathfrak{a}$ from elements of $L(-\mathfrak{B})$. The bilinear algorithm of multiplication in $k_{\mathfrak{a}}$ can be represented as a bilinear algorithm computing $(\Sigma_{i=1}^n x_i f_i)(\Sigma_{j=1}^n y_j f_j) \bmod \mathfrak{a}$ in the linear span of $L(-\mathfrak{B})\cdot L(-\mathfrak{B}) \subseteq L(-2\mathfrak{B})$ over k. By the choice of $f_1,\ldots,f_n$ we have $(\Sigma_{i=1}^n x_i f_i)(\Sigma_{j=1}^n y_j f_j) = \Sigma_{m=1}^n z_m f_m \bmod \mathfrak{a}$, where z_m are bilinear forms in $\bar{x} = (x_1,\ldots,x_n), \bar{y} = (y_1,\ldots,y_n)$ with coefficients from k. To reconstruct z_m uniquely we look at $f_i f_j$ as elements of $L(-2\mathfrak{B})$. If $g_1,\ldots,g_t$ is a basis of $L(-2\mathfrak{B})$ over k, $t = \dim_k L(-2\mathfrak{B})$, then we have $f_i f_j = \Sigma_{r=1}^t B_{ij}^r g_r$ for scalars $B_{ij}^r \in k$. Since the mapping $j: L(-\mathfrak{B}) \to K \to k_{\mathfrak{a}}$ is surjective we have $g_r \equiv \Sigma_{m=1}^n C_r^m f_m \bmod \mathfrak{a}$ for (scalars) $C_r^m \in k$. If we have a bilinear algorithm over k with multiplicative complexity μ representing the system of t bilinear forms $Z_r = \Sigma_{i,j=1}^n B_{ij}^r x_i y_j$: $r = 1,\ldots,t$, then we have a bilinear algorithm over k of the same multiplicative complexity, μ, representing z_m, because $z_m = \Sigma_{r=1}^t C_r^m Z_r$ $(m = 1,\ldots,n)$. To determine Z_r we look at values of g_r at $\mathcal{P}$ from $\mathfrak{D}$. By definition, for every $X \in K$, $X(\mathcal{P}) \in k_{\mathcal{P}} \cup \{\infty\}$, where $X(p) = \infty$ if $v_{\mathcal{P}}(X) < 0$ (i.e. X has a pole at $\mathcal{P}$) and $X(p) \in k_{\mathcal{P}}^*$ if $v_{\mathcal{P}}(X) = 0$. We form a $\mathrm{Card}(\mathfrak{D}) \times t$-matrix $A = (g_r(\mathcal{P}))$ for $\mathcal{P} \in \mathfrak{D}$ and $r = 1,\ldots,t$ of values of the basis of $L(-2\mathfrak{B})$ at prime divisors from $\mathfrak{D}$. We claim that the rank of this matrix over k is exactly t. Indeed, if this matrix has rank $< t$, then there are $\lambda_1,\ldots,\lambda_t$ from k, not all zero, such that $\Sigma_{r=1}^t \lambda_r g_r(\mathcal{P}) = 0$ for all $\mathcal{P} \in \mathfrak{D}$. This means that the function $X = \Sigma_{r=1}^t \lambda_r g_r \in K^*$ from $L(-2\mathfrak{B})$ has zeroes at all $\mathcal{P} \in \mathfrak{D}$. Since the degree of the divisor (X) is zero for $X \in K^*$, and $X \in L(-2\mathfrak{B})$, we have $\mathrm{Card}(\mathfrak{D}) < 2d(\mathfrak{B})$, which is impossible. Thus the matrix A has rank t. (Strictly speaking, this argument is valid only when the supports of $\mathfrak{D}$ and $\mathfrak{B}$ are disjoint; simple analysis show that this assumption can be removed by consideration of the intersection of $\mathfrak{D}$ and $\mathfrak{M}$.)

Let us consider a $t \times t$ submatrix A_0 of A, which is nonsingular; let its columns correspond to divisors $\theta_1,\dots,\theta_t$ of $\mathfrak{D}$. Then there exists a $t \times t$ matrix B_0 with elements from k such that $A_0B_0 = I_t$ in $M_t(k)$. We are ready to present now a bilinear algorithm over k of multiplicative complexity t (i.e. not more than $\mathrm{Card}(\mathfrak{D})$) which computes the bilinear forms Z_r: $r = 1,\dots,t$. First we make t linear forms in variables $\bar{x}$ and in the variable $\bar{y}$ separately:

$$X_s = \Sigma_{i=1}^{n} x_i f_i(\theta_s),\ Y_s = \Sigma_{j=1}^{n} y_j f_j(\theta_s)$$

$s = 1,\dots,t$. Then we form linear combinations of pairwise products of X_s and Y_s:

$$W_r = \sum_{s=1}^{t} B^0_{rs} X_s Y_s : r = 1,\dots,t$$

where $(B^0_{r,s})^t_{r,s=1} = B_0$. We claim that the bilinear forms W_r: $r = 1,\dots,t$ so defined coincide with the linear forms Z_r: $r = 1,\dots,t$. Indeed, by the definition of Z_r we have

$$\sum_{r=1}^{t} Z_r g_r = (\sum_{i=1}^{n} x_i f_i)(\sum_{j=1}^{n} y_j f_j),$$

i.e.

$$\sum_{r=1}^{t} Z_r g_r(\theta_s) = X_s Y_s : s = 1,\dots,t.$$

Consequently, for any $\alpha = 1,\dots,t$:

$$\sum_{s=1}^{t} B^0_{\alpha s} \sum_{r=1}^{t} Z_r g_r(\theta_s) = \sum_{s=1}^{t} B^0_{\alpha s} X_s Y_s,$$

or

$$\sum_{r=1}^{t} Z_r \sum_{s=1}^{t} B^0_{\alpha s} g_r(\theta_s) = \sum_{s=1}^{t} B^0_{\alpha s} X_s Y_s.$$

However $A_0 = (g_r(\theta_s))^t_{s,r=1}$ and $B_0A_0 = I_t$, i.e. $\Sigma^t_{s=1} B^0_{\alpha s} g_r(\theta_s) = \delta_{\alpha r}$. Thus for $\alpha = 1,\dots,t$

$$Z_\alpha = \sum_{s=1}^{t} B^0_{\alpha s} X_s Y_s \quad (= W_\alpha).$$

Consequently, we can determine Z_r: $r = 1,\dots,t$ in t essential multiplications over k and, as a consequence z_m: $m = 1,\dots,n$ can be determined over k in t essential multiplications too.

Corollary 5.2: Let K be a function field over k of genus $g \geq 0$ and let $\mathfrak{a}$ be a prime divisor of degree $n \geq 1$ on K. Let $\mathfrak{B}_0$ be a nonspecial integral divisor on K, i.e. $\dim_k L(-\mathfrak{B}_0) = d(\mathfrak{B}_0) - g + 1$, such that $\mathfrak{B} = \mathfrak{B}_0 + \mathfrak{a}$ is a nonspecial divisor too. Let there be D prime divisors of first degree on K for $D > 2d(\mathfrak{B})$. Then there exists a bilinear algorithm over k that computes the multiplication in the field extension $k_{\mathfrak{a}}$ of k of degree n of multiplicative complexity at most $2d(\mathfrak{B}) - g + 1 = 2n + 2d(\mathfrak{B}_0) - g + 1$.

Proof: According to Proposition 5.1 we have to prove that there exists an integral divisor $\mathfrak{B}_1$ of degree $d(\mathfrak{B})$ such that the mapping $j_{\mathfrak{B}_1}: L(-\mathfrak{B}_1) \to k_{\mathfrak{a}}$ is surjective. Since $j_{\mathfrak{B}_1}$ is determined by the place mapping $K \to k_{\mathfrak{a}}$ of $\mathfrak{a}$, for $j_{\mathfrak{B}_1}$ to map $L(-\mathfrak{B}_1)$ into $k_{\mathfrak{a}}$ one has to assume that supports of $\mathfrak{B}_1$ and $\mathfrak{a}$ are disjoint, i.e. that $\mathfrak{a}$ does not divide $\mathfrak{B}_1$. If $\mathfrak{B}$ is not divisible by $\mathfrak{a}$, we can choose $\mathfrak{B}_1 = \mathfrak{B}$. Otherwise we have to "move $\mathfrak{B}$ away from $\mathfrak{a}$", i.e. to choose $\mathfrak{B}_1$ as an integral divisor equivalent to $\mathfrak{B}$ not dividing $\mathfrak{a}$. Let C_0 be a class containing $\mathfrak{B}_0$, and C be a class containing $\mathfrak{B} = \mathfrak{B}_0 + \mathfrak{a}$, i.e. $C_0 = C - \mathfrak{a}$. If all integral divisors of the class C are divisible by $\mathfrak{a}$, then $N(C) = N(C-\mathfrak{a}) = N(C_0)$. However, by the assumption of the nonspeciality of $\mathfrak{B}_0$ and $\mathfrak{B}$, $N(C) = d(\mathfrak{B}) - g + 1 = d(\mathfrak{B}_0) + n - g + 1$ and $N(C_0) = d(\mathfrak{B}_0) - g + 1$. Consequently, $N(C) = N(C_0)$ is impossible for $n \geq 1$, and there is always an integral divisor $\mathfrak{B}_1$ in the class C, not divisible by $\mathfrak{a}$. We show now that the mapping $j_{\mathfrak{B}_1}: L(-\mathfrak{B}_1) \to k_{\mathfrak{a}}$ is surjective. For this we have to show that the kernel of $j_{\mathfrak{B}_1}$ has dimension over k: $\dim_k L(-\mathfrak{B}_1) - d(\mathfrak{a}) = N(C) - n$. This kernel is, on the other hand, $L(\mathfrak{a} - \mathfrak{B}_1)$ whose dimension over k is, by the assumption of the nonspeciality of $\mathfrak{B}_0$ (or C_0): $\dim_k L(\mathfrak{a} - \mathfrak{B}_1) = N(C-\mathfrak{a}) = N(C_0) = d(\mathfrak{B}_0) - g + 1$. The nonspeciality of $\mathfrak{B}$ (or C) implies, on the other hand, that $N(C) - n = d(\mathfrak{B}_1) - n - g + 1 = (n + d(\mathfrak{B}_0)) - n - g + 1 = d(\mathfrak{B}_0) - g + 1$. As a consequence, $j_{\mathfrak{B}_1}$ is surjective.

Let $\mathfrak{a}$ be an arbitrary prime divisor on K of degree n, and $k_{\mathfrak{a}}$ be its residue class field which is an extension of k of degree n. Let us denote by $\#K(k)$ the number of first degree divisors on K over k. It follows from Corollary 5.2 that whenever there exists a nonspecial divisor $\mathfrak{B}_0$ on K of degree m such that $m + n \geq 2g - 1$ and $\#K(k) \geq 2m + 2n$, the multiplicative complexity, $\delta_k(k_{\mathfrak{a}})$, of computation of multiplication in k_G over k, does not exceed $2m + 2n - g + 1$.

To prove the existence of the divisors necessary in Corollary 5.2, we use the properties of the ζ-function $Z(u)$ from (5.3)-(5.5). Let us denote by Nprime_n the number of prime divisors on K/k of degree n. It follows from (5.4) that $d/du \log Z(u) = \Sigma_{n=1}^{\infty} u^{n-1}\{\Sigma_{d|n}\mathrm{Nprime}_d \cdot d\}$. Comparing this expansion with the representation (5.5) for $P_{2g}(u) = \prod_{i=1}^{2g}(1-u\omega_i)$, we obtain:

$$q^n + 1 - \sum_{i=1}^{2g} \omega_i^n = \sum_{d|n} \mathrm{Nprime}_d \cdot d$$

for any $n \geq 1$. Since $|\omega_i| = \sqrt{q}$ for all $i = 1,\ldots,2g$, we deduce the following lower bound:

$$(5.6) \qquad n\cdot \mathrm{Nprime}_n \geq q^n - 2gq^{n/2} - q^{n/2+1} - 2gq^{n/4+1/2}.$$

From the bound (5.6) we deduce the following corollaries:

a) for every $n \geq c_1 \cdot \log(g)/\log q$ there is a prime divisor of degree n on K;

b) for every $m \geq g + c_2 \log(g.q)$ there exists a nonspecial positive divisor of degree m on K.

In order to have low multiplicative complexity algorithms arising from algebraic curves one sees that the remaining problem is the construction of algebraic curves over a fixed field $k = \mathbf{F}_q$ having the maximal number of divisors of the first degree for a given genus g, as $g \to \infty$.

The most general bound on the number $\#K(k)$ of points on K over k (prime divisors of first degree) is given by Weil's bound: $|N_k(K)-q-1| \leq 2g\sqrt{q}$. Unfortunately, as $g \to \infty$, the upper bound in the Weil theorem is unattainable. In fact, relatively simple considerations (cf. [49]) show that

$$(5.7) \qquad \lim_{g\to\infty} \frac{N_k(K)}{g} \leq \sqrt{g} - 1.$$

On the other hand, there are positive results [49], [50].

<u>Proposition 5.3</u>: Whenever q is square, the equality in (5.7) is attainable.

For example, on every classical modular curve of level m and genus g, $(m,p) = 1$, there are at least $(p-1)(g-1)$ points over $\mathbb{F}_{p^2}$.

Choosing curves over finite fields with many points on them (Fermat curves, other curves with complex multiplication, congruence curves) we obtain a variety of new low multiplicative complexity algorithms of polynomial multiplication.

Theorem 5.4: Let q be a square ≥ 25. Then the multiplicative complexity $\delta_{\mathbb{F}_q}(\mathbb{F}_{q^n})$ of multiplication in the field $\mathbb{F}_{q^n}$ over $\mathbb{F}_q$ can be bounded as follows:

$$\delta_{\mathbb{F}_q}(\mathbb{F}_{q^n}) \leq n\cdot 2\cdot(1 + \frac{1}{\sqrt{q}-3}) + o(n)$$

as $n \to \infty$.

Proof: We choose as a function field K over $k = \mathbb{F}_q$ a function field of a congruence curve Γ from Proposition 5.3 with $g \to \infty$. Then there are $D = g(\sqrt{q} - 1) + o(g)$ prime divisors of the first degree on K/k. We choose a nonspecial divisor $\mathfrak{B}_0$ of degree $m = g + o(g) \geq g$ on K/k and a prime divisor $\mathfrak{Q}$ of degree $n \geq g$ on K/k. Then we can identify $\mathbb{F}_{q^n}$ with the residue class field $k_{\mathfrak{Q}}$. The divisor $\mathfrak{Q} + \mathfrak{B}_0$ is nonspecial too because of the Riemann-Roch theorem (it has degree $\geq 2g$). According to Corollary 5.2 we have a bilinear algorithm over k that computes multiplications in $k_{\mathfrak{Q}}$ in $2(m+n) - g + 1$ nonscalar multiplications over k, whenever $D > 2(m+n)$, or $\sqrt{q} - 1 \geq 4$ (as $g \to \infty$). We finally choose g for a given n as $g = n\cdot 2/(\sqrt{q} - 3) + o(n)$; $n \geq g$. This gives the bound for $\delta_{\mathbb{F}_q}(\mathbb{F}_{q^n})$ as $\delta_{\mathbb{F}_q}(\mathbb{F}_{q^n}) \leq n\cdot 2(1+1/(\sqrt{q}-3)) + o(n)$.

For $q = 2$ we have the following upper and lower linear bounds on multiplicative complexities (see §4 and the multiplication rule (5.1)):

$$3.52\cdot n \leq \delta_{\mathbb{F}_2}(\mathbb{F}_{2^n}) \leq 6n \tag{5.8}$$

for a large n.

For $\mathbb{F}_2$ and moderate n and m (below 100) the best algorithm of polynomial multiplication over $\mathbb{F}_2$ has multiplicative complexity of linear character with a constant around 4.

Not all minimally complex algorithms arise from optimal linear codes. E.g. the minimal complexity over $\mathbb{F}_2$ of multiplication in the degree 4 field $\mathbb{F}_{2^4}$ over $\mathbb{F}_2$ is $\underline{9}$, while there is a Reed-Muller code [8,4,4] over $\mathbb{F}_2$. However, to prove that the minimal complexity is 9 we used the properties of the Reed-Muller code.

All algorithms of Theorem 5.4 and (5.8) are constructive (one can construct appropriate algebraic curve codes in polynomial time).

The construction of fast polynomial multiplication algorithms over $\mathbb{Z}$ is even harder than over $\mathbb{F}_2$. Several interesting problems of algebraic number theory immediately arise, when one attempts to construct these algorithms. One of them is the problem of exceptional units and Euclidean fields.

Apparently, one of the conditions that guarantees the existence of fast polynomial multiplications over $\mathbb{Z}$, is the existence of algebraic number fields K, $[K:\mathbb{Q}] = n$ with a (large) number M of "exceptional units". We say that $\varepsilon_1,\dots,\varepsilon_M$ are "exceptional units" in K if $\varepsilon_1 = 1$ and $\varepsilon_i - \varepsilon_j$ is a unit in K for $i \neq j$. This concept first appeared in works of Minkowski and Hurwitz. Recently it was studied by Lenstra [51]. In particular, Lenstra had shown that the following explicit bound

$$M > \frac{n!}{n^n}(4/\pi)^s \cdot |D_K|^{1/2}$$

(where D_K is the discriminant of K and s is the number of complex archimedian primes) on the number M of exceptional units implies the Euclideanity of K with respect to the norm $N_{K/\mathbb{Q}}(\cdot)$ of K.

Obvious bounds on M for an arbitrary field K of degree n are

$$2 \leq M \leq 2^n.$$

In fact, an upper bound on M is inflated for large n. It is possible to show that the bound is polynomial in n:

$$M \leq c_1 \cdot n^{c_2}$$

for $c_2 > 1$. On the other hand, by looking at subfields of cyclotomic fields and subfields of Abelian extensions of quadratic imaginary fields, one can construct infinite families of fields K of degree n such that

$$M = O(n \text{ loglog } n) \text{ and}$$

$$M = O(n \log^2 \log n)$$

for the number M of exceptional units in K.

As to the best algorithms of polynomial multiplication using only integer arithmetic, for large n we constructed algorithms that are asymptotically better than any algorithms arising from the fast discrete Fourier transform. For small n $(n \leq 100)$ there are, in fact, much faster methods, matching those over $\mathbb{F}_2$. For arbitrary n, we can improve the bound of the minimal complexity, $\mu_{\mathbb{Z}}(n,n)$ of multiplication of polynomials of degree $n - 1$ with the ring $\mathbb{Z}$ of constants by a factor $\log^2 \log\log n$ comparing to the best FFT bound $\mu_{\mathbb{Z}}(n,n) = O(n \log n)$.

§6. Multidimensional Laws of Addition for θ-functions.

We begin with a little history of some well-known identities. Halphen [52] showed that the famous Jacobi three-term identity on Weierstrass σ-functions (see [53]):

$$\sum_{a,b,c} \sigma(u+a)\sigma(u-a)\sigma(b+c)\sigma(b-c) = 0 \tag{6.1}$$

uniquely determines the σ-function or its degenerations (up to an exponential factor). The identity (6.1), or equivalent "parallelogram" identity

$$\frac{\sigma(u+v)\sigma(u-v)}{\sigma(u)^2\sigma(v)^2} = \wp(v) - \wp(u), \tag{6.2}$$

are sufficient to construct all (algebraic parts of) the theory of elliptic functions: see Halphen and Weierstrass [54]. In this theory, the Weierstrass elliptic functions $\wp(u)$ is defined as a function in the right hand side of (6.2); it turns out that with the proper norming of σ, one gets $\wp(u) = -(d/du)^2 \log \sigma(u)$ and, eventually, one recovers the full spectrum of elliptic function identities.

Weierstrass [54] suggested that the theory of θ and Abelian functions can be a priori deduced for $g > 1$ from identities similar to (6.1) that he conjectured. Such identities were indeed proved later by Caspary [55] and Frobenius [56]. These identities can be represented in the form

(6.3)
$$\sum_{\{1,3,5,\ldots,r+1\}} \sigma(\bar u_0+\bar u_1)\sigma(\bar u_0-\bar u_1)\sigma(\bar u_2+\bar u_3)\sigma(\bar u_2-\bar u_3)\cdots$$
$$\cdots\ \sigma(\bar u_r+\bar u_{r+1})\sigma(\bar u_r-\bar u_{t+1}) = 0.$$

Here $r = 2^g$, where $\sigma(\bar u)$ is an odd g-dimensional θ-function (see definitions in [57] or [58]). For example, one can take in usual notations $\sigma(u) = \theta[\frac{1}{2}\binom{m'}{m}](u)$ for odd characteristics $\frac{1}{2}\binom{m'}{m}$.

A variety of identities of this form (for θ-functions with different characteristic) was derived by Frobenius [56]. Some of these identities include less than $2^g + 2$ unknowns, as in (6.3), though one cannot have an identity of the form (6.3) for general θ-functions for any $r < 2^g$. E.g. in the case $g = 2$ one gets a six-term identity in four variables (and, in fact, there are 15 such identities):

(6.4)
$$\sum_{a,b,c} \wp[A](u+a)\wp[A](u-a)\wp[A](b+c)\wp[A](b-c)$$
$$= -e^{\pi i|AB|}\cdot \sum_{a,b,c} \wp[B](u+a)\wp[B](u-a)\wp[B](b+c)\wp[B](b-c),$$

where A and B are any two odd characteristics.

The existence of identities (6.3) and (6.4) is related to the generalization of the parallelogram identity (6.2). These new identities allow one to define a set of Abelian functions that are second logarithmic derivatives of the original σ-function. E.g. with the proper norming for $g = 2$ of the σ-function (at $\bar u = \bar 0$), Baker [58] deduced the representation

$$\frac{\sigma(\bar u+\bar v)\sigma(\bar u-\bar v)}{\sigma(\bar u)^2\sigma(\bar v)^2} = [\wp_{11}(\bar u)-\wp_{11}(\bar v)]$$
$$+ [\wp_{12}(\bar u)\wp_{22}(\bar v) - \wp_{12}(\bar v)\wp_{22}(\bar u)],$$

where

(6.6)
$$\wp_{ij}(\bar u) \stackrel{\text{def}}{=} -\frac{d}{du_i}\frac{d}{du_j}\log\sigma(\bar u).$$

Similarly, one can compute for $g = 3$ that in notations (6.6), for a properly normalized $\sigma(\bar u)$, one gets

$$\frac{\sigma(\bar{u}+\bar{v})\sigma(\bar{u}-\bar{v})}{\sigma(\bar{u})^2\sigma(\bar{v})^2} = \{\theta_{31}(\bar{u})-\theta_{31}(\bar{v})\}^2 - \{\theta_{33}(\bar{u})-\theta_{33}(\bar{v})\}\cdot\{\theta_{11}(\bar{u})-\theta_{11}(\bar{v})\}$$
$$+ \{\theta_{21}(\bar{u})-\theta_{21}(\bar{v})\}\cdot\{\theta_{23}(\bar{u})-\theta_{23}(\bar{v})\}$$
$$- \{\theta_{22}(\bar{u})-\theta_{22}(\bar{v})\}\cdot\{\theta_{31}(\bar{u})-\theta_{31}(\bar{v})\}.$$

In fact, the generalization of the parallelogram identity (6.2) becomes now the statement that $\frac{\sigma(\bar{u}+\bar{v})\sigma(\bar{u}-\bar{v})}{\sigma(\bar{u})^2\sigma(\bar{v})^2}$ is expressed as a sum of k products $(F(u)-F(v))\cdot(G(u)-G(v))$ and $F(u)\cdot G(v) - F(v)\cdot G(u)$ (where k is connected with the dimension g), instead of the simple form $F(\bar{u}) - F(\bar{v})$ of (6.2).

The appearance of these product terms (as compared to expressions for $\prod_{i=1}^{g} \frac{\sigma_i(u_i+v_i)\sigma_i(u_i-v_i)}{\sigma(u_i)^2\sigma(v_i)^2}$ represent the "obstructions" for the factorization equation of S-matrices expressed in terms of g-dimensional theta-functions. We refer readers to [59] for definitions and expressions of these S-matrices.

§7. Elliptic genera and modular forms of level 2.

One of the interesting aspects of formal groups associated with elliptic curves is their appearance in the classification of spin-manifolds with (odd) action of S^1. We describe briefly here our experience with this subject resulting in the derivation in the summer of 1985 of the formula for the characteristic power series of elliptic genera in terms of modular functions of level two (ratios of θ-functions). This formula was derived in [61] to answer a problem of Landweber-Stong [60] on integrality of coefficients of elliptic multiplicative characteristic class.

We adopt the standard definitions of [62]. By a genus one means a ring homomorphism $\varphi: \Omega_*^{SO} \to \Lambda$ from the oriented bordism ring to a commutative $\mathbb{Q}$-algebra with unit. The "universal" elliptic genus $\rho_t: \Omega_*^{SO} \to \mathbb{Q}[[t]]$, vanishing on all $[\mathbb{CP}(\xi^{2m})]$, arises from a multiplicative characteristic class $\rho_t = \Sigma_{k\geq 0}\rho_k t^k$ with $\rho_k = \pi_k + \Sigma_{0<|\omega|<k} a^k_\omega s^\pi_\omega$. One studies ρ_t by looking at the characteristic power series

(7.1) $$f_t(y) = 1 + yt + \Sigma_{k\geq 2}\, p_k(y)t^k,$$

where $p_k(y) = \Sigma_{i=1}^{k-1} a^k_{(i)}\, y^i \in \mathbb{Q}[y]$, $a^k_{(1)} = 0$: $k \geq 2$.

At the request of Landweber and Stong we studied the question of integrality of the coefficients of power series expansions of $f_t(y)$. In the course of the study of this question by means of IBM's SCRATCHPAD computer algebra system we derived a closed form expression for $f_t(y)$ in terms of modular functions.

To make a long story short, one looks at an elliptic curve $y^2 = 1 - 2\delta z^2 + \epsilon z^4$, where $\delta = \delta(t)$ and $\epsilon = \epsilon(t)$ are power series from $\mathbb{Q}[[t]]$, and at associated Jacobi's elliptic function $a \cdot \mathrm{sn}(\frac{x}{a}|k^2)$, where a and the modulus k^2 are connected with δ and ϵ as follows:

$$(7.2) \qquad a^2\delta = (1+k^2)/2, \; a^4\epsilon = k^2.$$

The relationship of $f_t(y)$ with $a \cdot \mathrm{sn}(\frac{x}{a}|k^2)$ is the following ([60],[61]):

$$(7.3) \qquad \frac{1}{a\mathrm{sn}(\frac{x}{a}|k^2)} = \frac{1}{2\sinh(\frac{x}{2})} f_t(e^x+e^{-x}-2).$$

From (7.3) (by looking at periods of functions in (7.3)) one deduces the fundamental relations between a and quater-periods K and K':

$$(7.4) \qquad \frac{2\pi i}{a} = 2K + 4mK + 2niK'$$

(for rational integers m and n). In [61] we determined the properties of relations (7.4) under the action of the modular group in the upper half plane of $\tau = i\frac{K'}{K}$. The knowledge of the leading term in the expansion of δ and ϵ in powers of t: $\delta \equiv -\frac{1}{8} + 3t$, $\epsilon = -t + 7t^2 + \ldots$ allows us to identify m and n in (7.4). Namely, we show that $1/a$ is represented as a power series in $\lambda = k^2$. Since $Y = 1/a$ satisfy, in view of (7.4), the Legendre's linear differential equation $[\lambda(1-\lambda)\frac{d^2}{d\lambda^2} + (1-2\lambda)\frac{d}{d\lambda} - \frac{1}{4}]Y = 0$, one can identify $1/a$ with K $(= \frac{\pi}{2}\,{}_2F_1(\frac{1}{2},\frac{1}{2};1,\lambda)$-- the basis of solutions regular at $\lambda = 0)$. From the modular structure of $1/a$ and its leading term in the expansion at $\lambda = 0$ we derive a single transcendental relation

$$(7.5) \qquad \frac{\pi i}{a} = K.$$

The relations (7.5) and (7.3) uniquely determine $f_t(y)$ as a function of $q = e^{\pi i\tau}$:

$$(7.6)\qquad f_t(y) = \prod_{n=1}^{\infty} \left[\frac{1-yq^{2n-1}/(1-q^{2n-1})^2}{1-yq^{2n}/(1-q^{2n})^2}\right].$$

The parameter t is related to q in the following way $t = -q + O(q^2)$. In fact, as it is noted in [61], the most natural choice of t is

$$t = -q,$$

which makes coefficients in the expansion of $f_t(y)$ in (7.6) into modular functions of τ.

We also considered in [61] normalization of $f_t(y)$ under the transformations $t \to t' = t + \Sigma_{k\geq 2} b_k t^k$. The most important normalization is that of $a^k_{(i)} = 0$ for $i > [\frac{k+1}{2}]$ in (7.1). In this case t becomes a modular function of q and can be represented in terms of ϑ-constants as follows:

$$(7.7)\qquad \begin{aligned} t &= \frac{1}{24}\{1 - \vartheta_2(0)^4 - \vartheta_3(0)^3\}, \text{ or} \\ t &= -\sum_{n=1}^{\infty} (2n+1)q^{2n+1}/(1-q^{2n+1}). \end{aligned}$$

As a consequence of the formulas, all coefficients in the expansion of $f_t(y)$ are rational integers.

Recently our formulas (7.6-7.7) were applied to some physical problems and reinterpreted in a geometric way by Landweber [64] and Witten [63]. In this interpretation one starts with characteristic classes Θ_t defined as follows. For (a real or complex) vector bundle E one put:

$$\lambda_t(E) = \sum_{n\geq 0} \lambda^n(E)t^n, \qquad S_t(E) = \sum_{n\geq 0} S^n(E)t^n,$$

where $\lambda^n(E)$, $S^n(E)$ are exterior and symmetric powers of E. Then Witten defines:

$$\Theta_t(E) = \bigotimes_{n=1}^{\infty} [\lambda_{t^{2n-1}}(E) \otimes S_{t^{2n}}(E)].$$

With this characteristic class one obtains a genus $\Theta_t : \Omega_*^{SO} \to \mathbb{Q}[[t]]$ in a natural way

$$\Theta_t(M^{4n}) = \hat{A}(M)\,\mathrm{ch}\,\frac{\Theta_t(TM)}{\Theta_t(1)^{4n}}[M^{4n}].$$

P. Landweber [64] verified that the genus Θ_t of Witten [63] coincides with the universal elliptic genus ρ_t of Landweber-Stong using our formula (7.6).

References.

[1] G. Pólya , G. Szegö, Aufgaben und Lehrsatze aus der Analysis, v. 2, Springer, N.Y., 1964.

[2] M. Hazewinkel, Formal Groups and Applications, Academic Press, 1978.

[3] P. Cartier, Groupes formels, fonctions automorphes et fonctions zéta des courbes elliptiques, Actes Congr. Int. Math. Nice, v. 2, 291-299, Gauthier-Villars, 1971.

[4] T. Honda, Formal groups and zeta functions, Osaka J. Math. 5 (1968), 199-213.

[5] T. Honda, On the theory of commutative formal groups, J. Math. Soc. Japan 22 (1970), 213-246.

[6] W. Hill, Formal groups and zeta functions of elliptic curves, Inv. Math. 12 (1971), 321-336.

[7] E. Ditters, Sur les congruences d'Atkin et de Swinnerton-Dyer, C.R. Acad. Sci. Paris, 282A (1976), 1131-1134.

[8] E. Ditters, The formal groups of an Abelian variety, defined over W(k), Vrije Univ. Amsterdam, Rep. 144, 1980.

[9] N. Yui, Formal groups and some arithmetic properties of elliptic curves, Lecture Notes Math. 732, Springer, 1979, 630-658.

[10] T. Honda, Algebraic differential equations, Symposia Mathematica v. 24, Academic Press, 1981, 169-204.

[11] C.L. Siegel, Über einige Anwendungen diophantischer Approximationen, Abh. Preuss. Akad. Wiss. Phys. Math. Kl. 1, 1929.

[12] D.V. Chudnovsky, G.V. Chudnovsky, Applications of Padé approximations to diophantine inequalities in values of G-functions, Lecture Notes Math. 1135, Springer, 1985 , 9-51.

[13] E. Bombieri, On G-functions, Recent Progress in Analytic Number Theory (ed. by M. Halberstam and C. Hooley), v. 2, Academic Press, 1981, 1-67.

[14] G.V. Chudnovsky, Measures of irrationality, transcendence and algebraic independence. Recent progress, Journées Arithmétiques 1980 (ed. by J.V. Armitage), Cambridge Univ. Press, 1982, 11-82.

[15] G. V. Chudnovsky, On applications of diophantine approximations, Proc. Natl. Acad. Sci. U.S.A., 81 (1984), 7261-7265.

[16] N. Katz, Nilpotent connections and the monodromy theorem, Publ. Math. I.H.E.S. 39 (1970), 355-412.

[17] N. Katz, Algebraic solutions of differential equations, Invent. Math. 18 (1972), 1-118.

[18] D.V. Chudnovsky, G.V. Chudnovsky, Applications of Padé approximations to the Grothendieck conjecture on linear differential equations, Lecture Notes Math. 1135, Springer, 1985, 52-100.

[19] D.V. Chudnovsky, G.V. Chudnovsky, Padé approximations and diophantine geometry, Proc. Natl. Acad. Sci. U.S.A. 82 (1985), 2212-2216.

[20] B. Dwork, Arithmetic theory of differential equations, Symposia Mathematica v. 24, Academic Press, 1981, 225-243.

[21] D.V. Chudnovsky, G.V. Chudnovsky, The Grothendieck conjecture and Padé approximations, Proc. Japan. Acad. 61A (1985), 87-91.

[22] C. Matthews, Some arithmetic problems on automorphisms of algebraic varieties, Number Theory Related to Fermat's Last Theorem, Birkhauser, 1982, 309-320.

[23] D.V. Chudnovsky, G.V. Chudnovsky, A random walk in higher arithmetic, Adv. Appl. Math. 7 (1986), 101-122.

[24] J. Tate, The arithmetic of elliptic curves, Invent. Math. 23 (1974), 179-206.

[25] G. Faltings, Endlichkeitssätze fur abelsche varietäten über zahlkörpern, Invent. Math. 73 (1983), 349-366.

[26] Y.I. Manin, Rational points on algebraic curves over function fields, Amer. Math. Soc. Translations (2) 50 (1966), 189-234.

[27] T. Honda, Two congruence properties of Legendre polynomials, Osaka J. Math., 13 (1976), 131-133.

[28] N. Yui, Jacobi quartics, Legendre polynomials and formal groups, this volume.

[29] J. Holt, On the irreducibility of Legendre's polynomials, II, Proc. London Math. Soc. (2) 12 (1913), 126-132.

[30] P. Landweber, Supersingular elliptic curves and congruences for Legendre polynomials, this volume.

[31] D.E. Knuth, The Art of Computer Programming, v. 2, 2nd ed., Addison-Wesley, 1981.

[32] J. Brillhart, D. Lehmer, J. Selfridge, New primality criteria and factorization of $2^m \pm 1$, Math.Comp. 29 (1975), 620-647.

[33] D. Chudnovsky, G. Chudnovsky, Sequences of numbers generated by addition in formal groups and new primality and factorization tests, Adv. Appl. Math. 7 (1986), 385-434.

[34] H. Lenstra, Algorithms of number theory, Proc. Int. Conference "Computers and Mathematics", Stanford, August 1986 (to appear).

[35] R. Schoof, Elliptic curves over finite fields and the computation of square roots mod p, Math. Comp. 44 (1985), 483-494.

[36] C. Fiduccia, I. Zalcstein, Algebras having linear multiplicative complexity, J. of ACM 24 (1977), 911-931.

[37] A. Adler, V. Strassen, On the algorithmic complexity of associative algebras, Theor. Comp. Sci. 15 (1981) 201-211.

[38] S. Winograd, Some bilinear forms whose multiplicative complexity depends on the field of constants, Math. Syst. Theory 10 (1977), 169-180.

[39] S. Winograd, Arithmetic Complexity of Computations, CBMS-NSF Regional Conf. Series Appl. Math., SIAM Publications #33, 1980.

[40] A. Schonhage, V. Strassen, Schnelle multiplikation grosser zahlen, Computing 7 (1971), 281-292.

[41] N. Nussbaumer, Fast Fourier Transform and Convolution Algorithms, Springer, 1982.

[42] F. MacWilliams, N. Sloane, The Theory of Error-Correcting Codes, North-Holland, Amsterdam, 1977.

[43] V. Goppa, Codes and information, Russian Math. Surveys 39 (1984), 87-141.

[44] H. de Groote, Characterization of division algebras of minimal rank and the structure of their algorithm varieties, SIAM J. Comput. 12 (1983), 101-117.

[45] R. McEliece, E. Rodenich, H. Rumsey, L. Welch, New upper bounds on the rate of a code via the Delsarte-MacWilliams inequalities, IEEE Trans.Inform. Theory 1T-23 (1977), 157-166.

[46] M. Brown, D. Dobkin, An improved lower bound on polynomial multiplication, IEEE Trans. Comp. 29 (1980), 237-240.

[47] M. Deuring, Lectures on the Theory of Algebraic Functions of One Variable, Springer, 1973.

[48] A. Weil, Courbes Algébriques et Variétés Abéliennes, Hermann, Paris, 1971.

[49] Y. Ihara, Some remarks on the number of rational points of algebraic curves over finite fields, J. Fac. Sci. Univ. Tokyo 28A (1981), 721-724.

[50] S. Vleduts, Y. Manin, Linear codes and modular curves, J. Soviet Math. 25 (1984), 2611-2643.

[51] K. Lenstra, Euclidean number fields of large degree, Invent. Math. 38 (1976/77), 237-254.

[52] G. Halphen, Traité des Fonctions Elliptiques et de Leurs Applications, Gauthier-Villars, V. 1-3, 1886-1890.

[53] E. Whittaker, G. Watson, Modern Analysis, Cambridge, 1927.

[54] K. Weierstrass, Zur Theorie der Jacobi 'schen Functionen von mehreren Veränderlichen, Sitzunber.Königl.Akad. Wissen. v. 4, 1882.

[55] F. Caspary, Ableitung des Weierstrasschen Fundamental-Theorems für die Sigmafunction mehrerer Argumente aus den Kroneckerschen Relationen für Subdeterminanten symmetrischer Systeme, J. Reine Angew. Math. 96 (1884), 182-184.

[56] G. Frobenius, Über Thetafunctionen mehrerer Variabeln, J. Reine Angew. Math. 96 (1884), 100-140.

[57] J.-I. Igusa, Theta Functions, Springer, 1972.

[58] N. Baker, Abel's Theorem and the Allied Theory Including the Theory of Theta Functions, Cambridge, 1890.

[59] D. Chudnovsky, G. Chudnovsky, Some remarks on theta function and S-matrices, Classical and Quantum Models and Arithmetic Problems, M. Dekker, 1986, 117-214.

[60] P. Landweber, R. Stong, Circle action on Spin manifolds and characteristic numbers, Topology (to appear).

[61] D. Chudnovsky, G. Chudnovsky, Elliptic modular functions and elliptic genera, Topology (to appear).

[62] F. Hirzebruch, Topological methods in algebraic geometry, Springer, 1966.

[63] E. Witten, Elliptic genera and quantum field theory, Commun. Math. Phys. 109 (1987), 525-536.

[64] P. Landweber, Elliptic cohomology and modular forms, this volume.

ELLIPTIC COHOMOLOGY AND MODULAR FORMS

by Peter S. Landweber

Rutgers University, New Brunswick, NJ 08903
Institute for Advanced Study, Princeton, NJ 08540

§1. Introduction. The homology and cohomology theories of the title, which have been found in joint work with Doug Ravenel and Bob Stong [14], are periodic complex-oriented multiplicative theories, with the cohomology of a point naturally interpreted as a ring of modular functions. The formal groups that occur for these theories are obtained from the formal group of the Jacobi quartic

$$Y^2 = 1 - 2\delta X^2 + \varepsilon X^4 \qquad (1)$$

over the ring $\mathbf{Z}[\frac{1}{2}][\delta,\varepsilon]$ by passing to suitable localizations of this ring, where δ and ε are viewed as indeterminates of degrees 4 and 8. We view these theories as belonging to a tower:

bordism and cobordism (MU, MSpin, MSO)
↓
elliptic cohomology (Ell)
↓
K-theory (KU, KO)
↓
ordinary cohomology (H)

In the first part of this report, I want to provide an assurance that such theories exist. In the second part, I shall explore the connections with modular forms.

There are several prominent open questions in this subject, the main one being to give a geometric definition of the elliptic cohomology theories. A number of these problems will be collected at the end.

It is a pleasure to thank the many people with whom I have discussed these topics; by now the list is extremely long. Thanks are also due to the National

Science Foundation and the Institute for Advanced Study for financial support.

§2. Elliptic genera. By a genus in the sense of Hirzebruch [7], one means a ring homomorphism

$$\varphi : \Omega_*^{SO} \to \Lambda$$

from the oriented bordism ring to a commutative $\mathbb{Q}$ - algebra with unit ($\varphi(1) = 1$) . Each such genus has a logarithm

$$g(x) = \int_0^x \sum_{n \geq 0} \varphi(\mathbb{C}P^{2n})\, t^{2n}\, dt ,$$

and a characteristic power series

$$u/g^{-1}(u) .$$

Following S. Ochanine [18], we call φ an elliptic genus if

$$g(x) = \int_0^x (1 - 2\delta t^2 + \varepsilon t^4)^{-\frac{1}{2}}\, dt \tag{2}$$

with elements $\delta, \varepsilon \in \Lambda$. In this case, the corresponding formal group

$$F(x,y) = g^{-1}(g(x) + g(y))$$

has the following form found by Euler:

$$F(x,y) = \frac{x\sqrt{R(y)} + y\sqrt{R(x)}}{1 - \varepsilon x^2 y^2} \tag{3}$$

where

$$R(x) = 1 - 2\delta x^2 + \varepsilon x^4 . \tag{4}$$

The signature (L-genus) and $\hat{A}$-genus are special cases. Namely, if $\delta = \varepsilon = 1$ then one has

$$g(x) = \int_0^x \frac{1}{1-t^2}\, dt = \tanh^{-1}(x)$$

and so obtains the characteristic series $u/\tanh u$ of the L-genus of Hirzebruch. And if $\delta = -1/8$, $\varepsilon = 0$ then one finds

$$g(x) = \int_0^x (1 + \tfrac{1}{4}t^2)^{-\frac{1}{2}}\, dt$$

and the characteristic series $u/2 \sinh (u/2)$ of the $\hat{A}$-genus.

We remark that for any elliptic genus φ one has

$$\delta = \varphi(\mathbb{C}P^2) , \quad \varepsilon = \varphi(\mathbb{H}P^2) . \tag{5}$$

Recalling that the Legendre polynomials $P_n(x)$ are defined by ([10])

$$(1 - 2xt + t^2)^{-\frac{1}{2}} = \sum_{n\geq 0} P_n(x)\, t^n ,$$

one sees easily that

(6) $$\varphi(\mathbb{CP}^{2n}) = P_n(\delta/\sqrt{\varepsilon})\, \varepsilon^{n/2} =: P_n(\delta,\varepsilon) .$$

It is a pleasant surprise that on quaternionic projective spaces one has ([5])

(7) $$\varphi(\mathbb{HP}^n) = \begin{cases} \varepsilon^{n/2} , & n \text{ even} \\ 0 , & n \text{ odd} . \end{cases}$$

In view of (3) and the binomial expansion, it is immediate that all coefficients of $F(x,y)$ are in $\mathbf{Z}[\frac{1}{2}]\,[\delta,\varepsilon]$, so by Quillen's theorem φ maps Ω_*^{SO} into the subring $\mathbf{Z}[\frac{1}{2}]\,[\delta,\varepsilon]$ of Λ (see §5 for more precise results).

For a Jacobi quartic (1) or the corresponding elliptic genus, we introduce the discriminant

(8) $$\Delta = \varepsilon(\delta^2 - \varepsilon)^2 .$$

If $\delta , \varepsilon \in \mathbb{C}$ and $\Delta \neq 0$, then $g^{-1}(u)$ is the expansion at the origin of an elliptic function $s(u)$, which is odd and of order 2 (a Jacobi sine; see [18] and §4). Note that the L-genus and $\hat{A}$-genus are "degenerate," i.e. $\Delta = 0$; the function $s(u)$ becomes singly periodic in these cases.

§3. <u>Elliptic homology and cohomology</u>. Continuing with the notation of the previous section, take δ and ε to be algebraically independent over $\mathbb{Q}$, and put

$$M_* = \mathbf{Z}[\tfrac{1}{2}]\,[\delta,\varepsilon] .$$

Then consider the rings:

$$\begin{array}{ccccc} & & M_*[\Delta^{-1}] & & \\ & \nearrow & & \nwarrow & \\ M_*[\varepsilon^{-1}] & & & & M_*[(\delta^2 - \varepsilon)^{-1}] \\ & \nwarrow & & \nearrow & \\ & & M_* & & \end{array}$$

<u>Theorem 1</u> ([14]). <u>There are homology theories with each of these rings as homology of a point. These are multiplicative theories, the corresponding cohomology theories being complex-oriented. The formal group of each of these theories has</u>

logarithm given by (2), and the explicit form of (3) with $R(x)$ as in (4).

Note: I am viewing homology and cohomology as two sides of the same coin. We write $Ell_*(X)$ and $Ell^*(X)$ for such theories.

First proof. We produce a connective homology theory with $Ell_*(pt) \cong M_*$, by using bordism with singularities (the Sullivan-Baas construction [1]). Thus start with oriented bordism theory with 2 inverted, $\Omega_*^{SO}(X)[\frac{1}{2}]$, a module over

$$\Omega_*^{SO}[\tfrac{1}{2}] = \mathbf{Z}[\tfrac{1}{2}][x_4, x_8, x_{12}, \ldots] .$$

We can take

$$x_4 = [\mathbb{C}P^2] , \; x_8 = [\mathbb{H}P^2]$$

and choose $x_{4n} = [M^{4n}]$ $(n \geq 3)$ so that the ideal

$$(x_{12}, x_{16}, \ldots)$$

consists of all bordism classes killed by elliptic genera; for the latter we follow Ochanine [18], generators for the ideal having the form $[\mathbb{C}P(\xi^{2m})]$ with ξ an even-dimensional complex vector bundle over a closed oriented manifold.

Now the Sullivan-Baas construction produces from the singularity set

$$\Sigma = \{x_{12}, x_{16}, \ldots\}$$

a theory $\Omega_*^{SO,\Sigma}[\frac{1}{2}](X)$ with $\Omega_*^{SO,\Sigma}[\frac{1}{2}](pt) \cong \mathbf{Z}[\frac{1}{2}][x_4, x_8] \xrightarrow{\sim} M_*$. Since 2 is inverted, one obtains a multiplicative homology theory, the obstructions to the existence of a good product all being 2-primary [16].

One can next simply invert Δ or its factors to obtain three further periodic theories.

Second proof. We shall follow a more insightful route, which yields the three periodic homology theories as a consequence of the exact functor theorem [12]. To explain the latter, let R be a commutative ring, and suppose given a formal group over R, i.e. a homomorphism from the complex bordism ring Ω_*^U to R (the formal group over Ω_*^U is universal). View R as a module over Ω_*^U. For each prime p and $n \geq 1$ define an element

$$u_n \in \Omega^U_{2(p^n-1)}$$

as the coefficient of z^{p^n} in the multiplication - by - p series

$$[p](z) = pz + \dots + u_1 z^p + \dots + u_2 z^{p^2} + \dots$$

for the universal formal group over Ω_*^U.

<u>Exact Functor Theorem</u> ([12]). <u>In order that</u>

$$X \to \Omega_*^U(X) \otimes_{\Omega_*^U} R$$

<u>be a homology theory, it suffices that for each prime</u> p

$$p, u_1, u_2, \dots, u_n, \dots$$

<u>be a regular sequence on the</u> Ω_*^U <u>- module</u> R. (I.e., multiplication by p on R and by each u_n on $R/(pR + \dots + u_{n-1}R)$ must be injective.)

Since we are inverting 2, it is the same to deal with

$$\Omega_*^{SO}(X) \otimes_{\Omega_*^{SO}} R$$

and apply the criterion for all <u>odd</u> primes. Here we take, say,

$$R = M_*[\Delta^{-1}] = \mathbf{Z}[\tfrac{1}{2}]\,[\delta, \varepsilon, \Delta^{-1}] .$$

With p an odd prime, multiplication by p on R is injective, and we pass to $\mathbb{F}_p[\delta, \varepsilon][\Delta^{-1}]$. In terms of the homogeneous Legendre polynomials of formula (6), one sees easily ([13]) that

(9) $$u_1 \equiv P_{(p-1)/2}(\delta, \varepsilon) \mod p .$$

That $u_1 \not\equiv 0 \mod p$ follows from the fact that $P_n(1) = 1$ for all n.

We are next obligated to examine $u_2 \mod (p, u_1)$, and here the principal facts are that

(10) $$\begin{cases} u_2 \equiv (-1)^{(p-1)/2}\, \varepsilon^{(p^2-1)/4} , \\ (\delta^2 - \varepsilon)^{(p^2-1)/4} \equiv \varepsilon^{(p^2-1)/4} \end{cases}$$

mod (p, u_1) in the ring $\mathbf{Z}[\tfrac{1}{2}]\,[\delta, \varepsilon]$. The point is that mod (p, u_1), inverting $\delta^2 - \varepsilon$ is equivalent to inverting ε, and so also to inverting Δ; and that u_2 then becomes a unit. This ends the argument for $R = M_*[\Delta^{-1}]$, and also in case just one factor of Δ is inverted. □

Note. The congruences (10) can be better appreciated if $\varepsilon = 1$, and then read

(11) $$\begin{cases} u_2 \equiv (-1)^{(p-1)/2} \\ (\delta^2 - 1)^{(p^2-1)/4} \equiv 1 \end{cases}$$

mod (p, u_1) in the ring $\mathbf{Z}[\frac{1}{2}][\delta]$. In this form, they were first pointed out by David and Gregory Chudnovsky [4]; the most direct proof is based on two papers of Igusa [8, 9]. Details will appear in [13], which also includes the following related result of Dick Gross.

Theorem ([6]). Let E be a supersingular elliptic curve given by a Weierstrass equation over a field of characteristic $p \geq 5$, so that for its formal group one has $[p](z) = u_2 z^{p^2} + \ldots$ with $u_2 \neq 0$, where $z = -x/y$ is the standard uniformizing parameter. Then

$$u_2 = (-1)^{(p-1)/2} \cdot \Delta^{(p^2-1)/12} ,$$

where Δ is the discriminant (expressed in terms of the coefficients of the Weierstrass equation).

§4. $\mathbf{Z}[\frac{1}{2}][\delta, \varepsilon]$ as a ring of modular forms. Let Γ denote the modular group $SL_2(\mathbf{Z})/\{\pm 1\}$, acting as usual on the upper half-plane H . Whereas Γ is generated by

$$\tau \mapsto \tau + 1 \quad , \quad \tau \mapsto -\tau^{-1} ,$$

it will be convenient here to deal with the subgroup Γ_θ generated by

$$\tau \mapsto \tau + 2 \; , \; \tau \mapsto -\tau^{-1} .$$

Γ_θ , the "theta group," has index 3 in Γ , and the standard fundamental domain with two cusps:

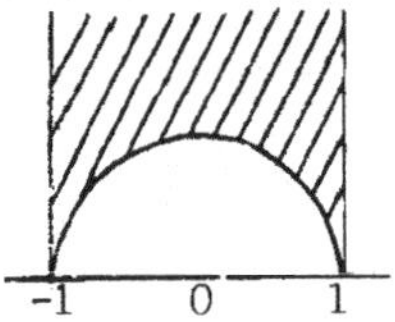

Moreover, $\begin{pmatrix} a & b \\ c & d \end{pmatrix} \in \Gamma_\theta$ if and only if $\begin{pmatrix} a & b \\ c & d \end{pmatrix}$ is congruent to $\begin{pmatrix} 1 & 0 \\ 0 & 1 \end{pmatrix}$ or $\begin{pmatrix} 0 & 1 \\ 1 & 0 \end{pmatrix}$ mod 2

The appearance of level 2 modular forms was first noticed by David and Gregory Chudnovsky [3].

In [13] I have found the following classical picture helpful. Assume δ, $\varepsilon \in \mathbb{C}$ and $\Delta \neq 0$, so

$$Y^2 = 1 - 2\delta X^2 + \varepsilon X^4$$

is an elliptic curve over $\mathbb{C}$. This curve is uniformized by an elliptic function $s(u)$, odd and of order 2, with period lattice generated by $2\omega_1$ and $2\omega_2$.

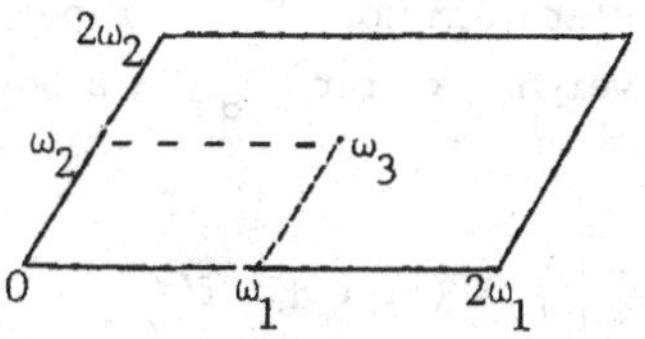

The function $s(u)$ has poles at ω_1 and ω_2, and zeros at 0 and $\omega_3 = \omega_1 + \omega_2$. Hence one of the half-periods ω_3 is distinguished; assume $\tau = \omega_2/\omega_1 \in H$. The lattice has the usual Weierstrass function $\wp(u)$ and invariants g_2, g_3, as well as the half-period values $e_i = \wp(\omega_i)$. The following formulas are now easily obtained:

$$s(u) = -2\,(\wp(u) - e_3)/\wp'(u)$$

$$\begin{cases} g_2 = (\delta^2 + 3\varepsilon)/3 \\ g_3 = \delta(\delta^2 - 9\varepsilon)/27 \end{cases}$$

$$\begin{cases} \delta = 3e_3 \\ \varepsilon = (e_1 - e_2)^2 \\ \delta^2 - \varepsilon = 4(e_1 - e_3)(e_2 - e_3) \end{cases}$$

Moreover, one can as a further exercise (see [2]) express all these quantities in terms of τ via theta functions. Taking $\omega_1 = \pi$ to remove powers of π/ω_1 from the expressions, one has

$$\begin{cases} \delta = \theta_1^4 - \theta_2^4 \\ \varepsilon = (\theta_1^4 + \theta_2^4)^2 = \theta_3^8 . \end{cases}$$

Here the "theta-constants" are given by $(q = e^{\pi i\tau})$

$$\theta_1(\tau) = 2q^{\frac{1}{4}} \sum_{n=0}^{\infty} q^{n(n+1)}$$

$$\theta_2(\tau) = 1 + 2 \sum_{n=1}^{\infty} (-1)^n q^{n^2}$$

$$\theta_3(\tau) = 1 + 2 \sum_{n=1}^{\infty} q^{n^2} .$$

One then finds easily that δ and ε are modular forms of weights 2 and 4 for Γ_θ, and indeed that every modular form for Γ_θ is a polynomial in δ and ε. Recall that a modular form of weight k for Γ_θ is a holomorphic function $f(\tau)$ on H so that

$$f\left(\frac{a\tau + b}{c\tau + d}\right) = (c\tau + d)^k f(\tau)$$

for $\begin{pmatrix} a & b \\ c & d \end{pmatrix} \in \Gamma_\theta$, and so that

$$f(\tau) = \sum_{n\geq 0} a_n q^n \qquad (q = e^{\pi i\tau})$$

with a similar holomorphicity condition at the other cusp. We see that

$$\delta(i\infty) = -1 , \quad \varepsilon(i\infty) = 1 ;$$

for the cusp at $\tau = 1$, we use $\tau \mapsto 1 - 1/\tau$ sending ∞ to 1 and find that

$$\delta(1) = 2 , \quad \varepsilon(1) = 0 .$$

We conclude that, up to inessential multiples, one finds the L-genus at $\tau = i\infty$ and the Â-genus (better, the A-genus) at $\tau = 1$.

Furthermore, in addition to

$$M_*(\Gamma_\theta) = \mathbb{C}[\delta , \varepsilon]$$

we can identify $M_* = \mathbf{Z}[\frac{1}{2}] [\delta , \varepsilon]$ with the ring of modular forms for Γ_θ with q-expansion coefficients in $\mathbf{Z}[\frac{1}{2}]$. In addition, we go on to identify $M_*[\Delta^{-1}]$ with those modular functions for Γ_θ which are holomorphic on H (poles at cusps only) and have q-expansion coefficients in $\mathbf{Z}[\frac{1}{2}]$. The rings $M_*[\varepsilon^{-1}]$ and $M_*[(\delta^2 - \varepsilon)^{-1}]$ are given similar interpretations, allowing a pole at one or the other cusp.

Thus we are to view δ and ε as modular forms, and so we should regard the elliptic genus

$$\varphi:\ \Omega_*^{SO} \to \mathbf{Z}[\tfrac{1}{2}]\ [\delta\ ,\ \varepsilon] = M_*$$

as assigning a modular form

$$\varphi(M^{4n}) = P_M(\delta\ ,\ \varepsilon)$$

to each oriented (or Spin) manifold. I leave it to Ed Witten [19] to explain a geometric procedure to produce such modular forms for Spin manifolds; we shall examine a formula for the resulting modular form in §6, given in terms of familiar constructions on vector bundles. In the next section, we answer the question: Which modular forms arise from oriented or from Spin manifolds?

§5. Integrality and divisibility of elliptic genera. The results stated here are taken from [5]. Let $\varphi:\ \Omega_*^{SO} \to \Lambda$ be an elliptic genus, with parameters $\delta\ ,\ \varepsilon \in \Lambda$.

Theorem 2 ([5]). For an elliptic genus, one has

$$\varphi\Omega_*^{SO} = \mathbf{Z}[\delta,\ 2\gamma,\ 2\gamma^2,\ldots,\ 2\gamma^{2^s},\ldots]$$

with $\gamma = (\delta^2 - \varepsilon)/4$, and

$$\varphi\Omega_*^{Spin} = \mathbf{Z}[16\delta,\ (8\delta)^2,\ \varepsilon]\ .$$

Corollary 1. For an elliptic genus $\varphi:\ \Omega_*^{SO} \to \mathbf{Q}$ one has

$$\varphi\Omega_*^{SO} = \mathbf{Z}[\delta\ ,\ \gamma]\ ,$$

$$\varphi\Omega_*^{Spin} = \mathbf{Z}[8\delta\ ,\ \varepsilon]\ .$$

Corollary 2 (Ochanine [17]). The signature of a Spin manifold M^{8k+4} is divisible by 16.

For the second corollary, take $\delta = \varepsilon = 1$ and note the degrees of the generators of $\varphi\Omega_*^{Spin}$.

We shall sketch the proof of the theorem for Ω_*^{Spin} .

Proof for Ω_*^{Spin} .

i) $\varphi(\mathbb{H}P^2) = \varepsilon$.

ii) There is a Spin manifold V^4 with signature 16, i.e. $\varphi(V^4) = 16\delta$.

iii) Kervaire and Milnor [11] constructed an almost parallelizable W^8 with $\hat{A}(W^8) = 1$; if

$$\varphi(W^8) = a\delta^2 + b\varepsilon$$

then $a = 64$; by i), we have $\varphi(M^8) = (8\delta)^2$ for a suitable Spin manifold.

iv) Hence $\mathbf{Z}[16\delta, (8\delta)^2, \varepsilon] \subset \varphi(\Omega_*^{Spin})$.

v) In [15] we constructed a sequence ρ_k , $k \geq 0$, of KO-theory characteristic classes of oriented bundles such that

$$\rho_k[M^{4n}] = \hat{A}(M)\, ch(\rho_k TM)\, [M^{4n}]$$

has the properties

a) $\rho_0[M^{4n}] = \hat{A}(M^{4n})$

b) $\rho_1[M^{4n}] = \hat{A}(M)\, ch(TM - 4n\mathbb{R})\, [M^{4n}]$

c) $\rho_k[M^{4n}]$ is integral on Spin manifolds, and is even when n is odd .

d) $\rho_t\colon \Omega_*^{SO} \to \mathbb{Q}[[t]]$ given by $\rho_t(M) = \sum_{k\geq 0} \rho_k[M]\, t^k$ is an elliptic genus,

for which

e) $\delta(t) \equiv -\frac{1}{8} + 3t \mod t^2\mathbf{Z}[[t]]$

f) $\varepsilon(t) \equiv -t \mod t^2\mathbf{Z}[[t]]$

(for the integrality in e) and f), see [3], [20] and the next section).

vi) Returning to the argument, if M^{8k} is a Spin manifold and

$$\varphi(M^{8k}) = a_0(8\delta)^{2k} + a_1(8\delta)^{2k-2}\varepsilon + \ldots + a_k\, \varepsilon^k$$

for an arbitrary elliptic genus φ , take $\varphi = \rho_t$ so that

$$\rho_t(M^{8k}) \in \mathbf{Z}[[t]]$$

to conclude easily that each $a_i \in \mathbf{Z}$. And if M^{8k+4} is a Spin manifold, write

$$\varphi(M^{8k+4}) = 16\delta \sum_{j=0}^{k} b_j (8\delta)^{2k-2j} \varepsilon^j$$

and argue that each $b_j \in \mathbf{Z}$. As an illustration of the simple method, if

$$\varphi(M^k) = 16\delta\ (b_0 (8\delta)^2 + b_1 \varepsilon)$$

then

$$16(-\frac{1}{8} + 3t + \ldots)\ [b_0(1 - 24t + \ldots)^2 + b_1(-\ t + \ldots)]$$

lies in $2\mathbf{Z}[[t]]$, so

$$b_0(1 - 24t + \ldots)^2 + b_1(1 - 24t + \ldots)\ (-\ t + \ldots)$$

lies in $\mathbf{Z}[[t]]$, whence

$$b_0 \in \mathbf{Z} \quad \text{(constant term)} ,$$

$$b_1 \in \mathbf{Z} \quad \text{(coefficient of } t) . \qquad \square$$

§6. <u>Witten's formula for the elliptic genus</u>. We refer to [19] for the geometry underlying the following considerations. For a real or complex vector bundle E , put

$$\lambda_t(E) = \sum_{k\geq 0} \lambda^k(E)\ t^k\ ,\ S_t(E) = \sum_{k\geq 0} S^k(E)\ t^k$$

where $\lambda^k(E)$ and $S^k(E)$ denote the exterior and symmetric powers of E . In addition, put

$$\Theta_t(E) = \bigotimes_{n=1}^{\infty} [\lambda_{t^{2n-1}}(E) \otimes S_{t^{2n}}(E)] .$$

Evidently, $\Theta_t(E \oplus F) = \Theta_t(E) \cdot \Theta_t(F)$. For a complex line bundle L we have

$$\Theta_t(L) = \prod_{n=1}^{\infty} \frac{1 + t^{2n-1}L}{1 - t^{2n}L} ,$$

in particular

$$\Theta_t(1) = \prod_{n=1}^{\infty} \frac{1 + t^{2n-1}}{1 - t^{2n}} .$$

One computes:

$$\Theta_t(E) = 1 + tE + t^2(\lambda^2 E + E) + t^3(\lambda^3 E + E \otimes E + E) + \dots$$

and immediately sees a connection with results of [15], indeed a step toward an explanation of these results.

Using this characteristic class, we obtain a genus

$$\Theta_t : \Omega_*^{SO} \to \mathbb{Q}[[t]]$$

by putting

$$\Theta_t(M^{4n}) = \hat{A}(M)\ \mathrm{ch}\ \frac{\Theta_t(TM)}{\Theta_t(1)^{4n}}\ [M^{4n}]$$

Theorem 3. The genus Θ_t is an elliptic genus, coinciding with the natural choice for the elliptic genus ρ_t .

After some clarifying remarks, one will see that this is really an easy observation. The analysis of elliptic genera ρ_t (see §5) in [15, 3] led to a formula for the characteristic power series

$$\frac{x/2}{\sinh(x/2)}\ f_t(e^x + e^{-x} - 2)\ .$$

Here

$$t \equiv -q \mod q^2\mathbf{Z}[[q]]\ ,$$

the natural parameter being q and the most natural choice of t being simply $t = -q$. In [3] (see also [20]) it is shown that, in terms of q ,

$$f_t(y) = \prod_{n=1}^{\infty} \frac{1 - y\, q^{2n-1}/(1 - q^{2n-1})^2}{1 - y\, q^{2n}/(1 - q^{2n})^2}\ .$$

Now the genus Θ_t is given in a form which permits one to easily find its characteristic power series. Indeed, with $t = -q$ as suggested above, one finds that for a complex line bundle L with $c_1(L) = x$ one has

$$\mathrm{ch}\ \Theta_t(L \oplus \bar{L} - 2)$$

$$= \prod_{n=1}^{\infty} \frac{(1 + t^{2n-1}e^x)(1 + t^{2n-1}e^{-x})/(1 + t^{2n-1})^2}{(1 - t^{2n}e^x)(1 - t^{2n}e^{-x})/(1 - t^{2n})^2}$$

$$= \prod_{n=1}^{\infty} \left[(1 - q^n e^x)(1 - q^n e^{-x})/(1 - q^n)^2\right]^{(-1)^{n-1}}$$

$$= \prod_{n=1}^{\infty} \left[1 - y\, q^n/(1 - q^n)^2\right]^{(-1)^{n-1}} ,$$

with $y = e^x + e^{-x} - 2$. We have therefore verified that

$$\mathrm{ch}\ \Theta_t(L \oplus \bar{L} - 2) = f_t(y) ,$$

whence the genera Θ_t and ρ_t (with $t = -q$) have identical characteristic series, the latter being an elliptic genus. The theorem is proved. □

Note. We refer to [20] for the interpretation of δ, ε, and so the elliptic genus of any oriented manifold, as modular forms for the group $\Gamma_0(2)$. This group is a conjugate of Γ_θ in Γ , consisting of all $\begin{pmatrix} a & b \\ c & d \end{pmatrix} \in \Gamma$ with c even.

7. Some open problems. As of October 1986, here are some rather naive questions that deserve study.

A) Give an intrinsic geometric construction for an elliptic cohomology theory.

B) Construct such a theory in which it is not necessary to invert 2.

C) Find appropriate versions of representation theory and index theory fitting with elliptic cohomology.

D) Since we are producing periodic cohomology theories, one might seek a fundamental result analogous to Bott periodicity, in an appropriate setting.

E) Spin bundles are orientable for elliptic cohomology. Construct such an orientation, compatibly with the KO-theory orientation constructed by means of Clifford algebras.

F) Seek related invariants in dimensions not divisible by 4. This may call for modular forms of half-integral weight, or mod p modular forms (especially with $p = 2$) .

G) Develop a variant in which modular forms of level 1 occur.

REFERENCES

1. N.A. Baas: On bordism theories of manifolds with singularities, Math. Scand. 33 (1973), 279-302.

2. K. Chandrasekharan: Elliptic Functions, Springer-Verlag, 1985.

3. D.V. Chudnovsky and G.V. Chudnovsky: Elliptic modular functions and elliptic genera, Topology, to appear.

4. D.V. Chudnovsky and G.V. Chudnovsky: letter dated February 6, 1986.

5. D.V. Chudnovsky, G.V. Chudnovsky, P.S. Landweber, S. Ochanine and R.E. Stong: Integrality and divisibility of elliptic genera, in preparation.

6. B.H. Gross: letter dated April 7, 1986.

7. F. Hirzebruch: Topological Methods in Algebraic Geometry, Springer-Verlag, 1966.

8. J. Igusa: On the transformation theory of elliptic functions, Amer. J. Math. 81 (1959), 436-452.

9. J. Igusa: On the algebraic theory of elliptic modular functions, J. Math. Soc. Japan 20 (1968), 96-106.

10. D. Jackson: Fourier Series and Orthogonal Polynomials, Math. Assoc. Amer., 1941.

11. M. Kervaire and J. Milnor: Bernoulli numbers, homotopy groups and a theorem of Rohlin, Proc. Int. Cong. Math., Edinburgh (1958), 454-458.

12. P.S. Landweber: Homological properties of comodules over MU_*MU and BP_*BP, Amer. J. Math. 98 (1976), 591-610.

13. P.S. Landweber: Supersingular elliptic curves and congruences for Legendre polynomials, in this volume.

14. P.S. Landweber, D.C. Ravenel and R.E. Stong: Periodic cohomology theories defined by elliptic curves, in preparation.

15. P.S. Landweber and R.E. Stong: Circle actions on Spin manifolds and characteristic numbers, Topology, to appear.

16. O.K. Mironov: Multiplications in cobordism theories with singularities, and Steenrod - tom Dieck operations, Izv. Akad. Nauk SSSR, Ser. Mat. 42 (1978), 789-806 = Math. USSR Izvestija 13 (1979), 89-106.

17. S. Ochanine: Signature modulo 16, invariants de Kervaire généralisés, et nombres caractéristiques dans la K-théorie réelle, Supplément au Bull. Soc. Math. France 109 (1981), Mémoire n° 5.

18. S. Ochanine: Sur les genres multiplicatifs définis par des intégrales elliptiques, Topology 26 (1987), 143-151.

19. E. Witten: Elliptic genera and quantum field theory, Communications in Mathematical Physics 109 (1987), 525-536.

20. D. Zagier: Note on the Landweber - Stong elliptic genus, in this volume.

SUPERSINGULAR ELLIPTIC CURVES AND CONGRUENCES FOR LEGENDRE POLYNOMIALS

by Peter S. Landweber

Rutgers University, New Brunswick, NJ 08903

Institute for Advanced Study, Princeton, NJ 08540

§1. Introduction. Recently, a number of applications of elliptic curves and classical elliptic function theory have been found in algebraic topology [2, 4, 5, 12, 13, 14, 17]. This has led to several questions concerning the formal groups of elliptic curves, which we shall examine here. These questions arose from the desire to gain a thorough understanding of the formal groups defined by the elliptic genera $\varphi\colon \Omega_*^{SO} \to R$ introduced by Ochanine [17], and in order to establish the existence of new periodic homology and cohomology theories related to elliptic curves and modular functions [12, 13].

An elliptic curve given by a Jacobi quartic

$$Y^2 = 1 - 2\delta X^2 + \varepsilon X^4 =: R(X) , \tag{1.1}$$

with uniformizer X near the origin $(X,Y) = (0,1)$, has invariant differential

$$dX/Y = (\sum_{n\geq 0} P_n(\delta,\varepsilon)\, X^{2n})\, dX$$

for suitable polynomials $P_n(\delta,\varepsilon)$. These polynomials are easily written out by means of the binomial series, and are "homogeneous Legendre polynomials." Indeed, one customarily defines the Legendre polynomials $P_n(x) \in \mathbf{Z}[\frac{1}{2}][x]$ by ([11, 22])

$$(1 - 2xt + t^2)^{-\frac{1}{2}} = \sum_{n\geq 0} P_n(x)\, t^n ,$$

from which it follows at once that

$$P_n(\delta,\varepsilon) = P_n(\delta/\sqrt{\varepsilon})\, \varepsilon^{n/2} ,$$

$$P_n(x,1) = P_n(x) .$$

The corresponding formal group was found explicitly by Euler ([7, 9, 16], see also the appendix):

$$F_E(X_1,X_2) = \frac{X_1\sqrt{R(X_2)} + X_2\sqrt{R(X_1)}}{1 - \varepsilon X_1^2 X_2^2} \tag{1.2}$$

where $R(X) = 1 - 2\delta X^2 + \varepsilon X^4$. If one views δ and ε as indeterminates, then $F_E(X_1,X_2)$ is a formal group over $\mathbf{Z}[\frac{1}{2}][\delta,\varepsilon]$.

As with any formal group, one has a multiplication-by-m series

$$[m](X) = mX + \text{higher terms}$$

for each integer m. For an odd prime p, one has

$$[p](X) = pX + \ldots + u_1X^p + \ldots + u_2X^{p^2} + \ldots$$

with (see §2 for details)

$$(1.3)\qquad \begin{cases} u_1 \equiv P_{(p-1)/2}(\delta,\varepsilon) \mod p \\ u_2 \equiv \frac{1}{p}\{P_{(p^2-1)/2}(\delta,\varepsilon) - P_{(p-1)/2}(\delta,\varepsilon)^{1+p}\} \mod (p,u_1)\,. \end{cases}$$

There is no loss in replacing δ by x and ε by 1; thus it is convenient to define polynomials $V_1(x)$ and $V_2(x)$ by

$$(1.3')\qquad \begin{cases} V_1(x) = P_{(p-1)/2}(x) \\ V_2(x) = \frac{1}{p}\{P_{(p^2-1)/2}(x) - P_{(p-1)/2}(x)^{1+p}\}\,; \end{cases}$$

indeed, $V_2(x)$ lies in $\mathbf{Z}[\frac{1}{2}][x]$ (see §2).

One had reason to inquire about $V_2(x)$ modulo the ideal $(p,V_1(x))$, and expected it to be a unit in

$$\mathbf{Z}[\tfrac{1}{2}][x]/(p,V_1(x))\,;$$

compare [12, 13]. Our main result, first proved by D. and G. Chudnovsky [3, 4] apart from (1.5'), answers this question; $(\frac{-1}{p})$ below is the usual Legendre symbol.

<u>Theorem 1</u>. <u>For an odd prime</u> p, <u>the polynomials</u> $V_1(x)$ <u>and</u> $V_2(x)$ <u>defined by</u> (1.3') <u>in terms of Legendre polynomials satisfy the congruences</u>

(1.4) $$V_2(x) \equiv \left(\frac{-1}{p}\right) = (-1)^{(p-1)/2} \mod (p,V_1(x))\,,$$

(1.5) $$(x^2-1)^{(p^2-1)/4} \equiv 1 \mod (p,V_1(x))\,, \text{ and}$$

(1.5') $$(x^2-1)^{(p^2-1)/12} \equiv 1 \mod (p,V_1(x)) \text{ if } p > 3\,.$$

For use in [12] and [13], we note an immediate consequence.

<u>Corollary</u>. <u>For an odd prime</u> p, <u>the following congruences hold in</u> $\mathbf{Z}[\delta,\varepsilon]$ mod $(p,P_{(p-1)/2}(\delta,\varepsilon))$:

$$\frac{1}{p}\{P_{(p^2-1)/2}(\delta,\varepsilon) - P_{(p-1)/2}(\delta,\varepsilon)^{1+p}\} \equiv (-1)^{(p-1)/2}\,\varepsilon^{(p^2-1)/4}\,,$$

$$(\delta^2-\varepsilon)^{(p^2-1)/4} \equiv \varepsilon^{(p^2-1)/4}\,.$$

Making use of (1.3), the first of these may be written more concisely as

$$u_2 \equiv (-1)^{(p-1)/2}\,\varepsilon^{(p^2-1)/4} \mod (p,u_1)\,.$$

We shall give two proofs of Theorem 1. In §2, it is shown to follow from corresponding results about supersingular elliptic curves (Theorems 2 and 3), which are established in sections 3 and 4. A more direct argument for (1.4) and (1.5) is given in §6, based on Igusa's studies of Jacobi quartics [9, 10].

In an attempt to understand the general phenomena underlying the congruences of Theorem 1, it appeared advisable to ask an analogous question about supersingular eliptic curves. Thus, let E be an elliptic curve over a field K of characteristic $p > 0$, with Weierstrass equation

$$y^2 + a_1xy + a_3y = x^3 + a_2x^2 + a_4x + a_6 \tag{1.6}$$

(see Silverman [20] for notation and standard facts about elliptic curves). Choose the standard uniformizing parameter $z = -x/y$ at the origin (at ∞) of E, and let $[p](z)$ be the power series giving multiplication by p in the formal group of E. Then in the expansion

$$[p](z) = pz + \ldots + u_1z^p + \ldots + u_2z^{p^2} + \ldots$$

one has $p = 0$ in K, and E is called <u>supersingular</u> if also $u_1 = 0$ in K. In the latter case, one knows that $u_2 \neq 0$ because the formal group of an elliptic curve over a field of finite characteristic has height 1 or 2 ([20, Cor. IV. 7.5]). In addition, the discriminant

$$\Delta = \Delta(a_1,\ldots,a_6)$$

is non-zero for any elliptic curve, and so it seemed reasonable to expect a simple relation between Δ and u_2 (of weights 12 and $p^2 - 1$, respectively) in the supersingular case. This was shown to be true by B.H. Gross [8]; we give his proof of the following result in §3.

<u>Theorem 2.</u> <u>For a supersingular elliptic curve, given by a Weierstrass equation</u> (1.6), <u>over a field of characteristic</u> $p > 0$, <u>the non-zero elements</u> Δ <u>and</u> u_2 <u>satisfy the equality</u>

$$u_2^4 = \Delta \quad \text{if } p = 2,$$

$$u_2^3 = -\Delta^2 \quad \text{if } p = 3, \text{ and}$$

$$u_2 = \left(\frac{-1}{p}\right)\Delta^{(p^2-1)/12} = (-1)^{(p-1)/2}\,\Delta^{(p^2-1)/12}$$

<u>if</u> $p > 3$.

The following supplementary information will be needed. Take an elliptic curve over a field of characteristic $p > 3$, given in the classical Weierstrass form

$$\begin{aligned} y^2 &= 4x^3 - g_2x - g_3 \\ &= 4(x - e_1)(x - e_2)(x - e_3) \end{aligned} \tag{1.8}$$

and having discriminant

$$\begin{aligned}\Delta &= g_2^3 - 27g_3^2 \\ &= 16(e_1 - e_2)^2 (e_1 - e_3)^2 (e_2 - e_3)^2\end{aligned}$$

Theorem 3. For a supersingular elliptic curve (1.8) over a field of characteristic $p > 3$, one has

(1.9) $$\Delta^{(p^2-1)/24} = (e_\alpha - e_\beta)^{(p^2-1)/4} \quad \text{if} \quad \alpha \neq \beta ,$$

and so in particular

(1.9') $$\Delta^{(p^2-1)/12} = (e_\alpha - e_\beta)^{(p^2-1)/2} \quad \text{if} \quad \alpha \neq \beta .$$

We shall combine Theorems 2 and 3 to prove Theorem 1, with the help of the following Theorem 4.

The conversion between elliptic curves in the Jacobi form

(1.1) $$Y^2 = 1 - 2\delta X^2 + \varepsilon X^4$$

with origin at $(X,Y) = (0,1)$, and in the Weierstrass form

(1.8) $$\begin{aligned}y^2 &= 4x^3 - g_2x - g_3 \\ &= 4(x - e_1)(x - e_2)(x - e_3)\end{aligned}$$

with origin at ∞, will be carried out as follows. Given elements δ and ε of a field K of characteristic different from 2 and 3, which satisfy

$$\Delta := \varepsilon(\delta^2 - \varepsilon)^2 \neq 0 ,$$

define g_2 and g_3 by

(1.10) $$\begin{cases} g_2 = (\delta^2 + 3\varepsilon)/3 , \\ g_3 = \delta(\delta^2 - 9\varepsilon)/27 . \end{cases}$$

One checks easily (see §5) that the equation $4x^3 - g_2x - g_3 = 0$ has roots e_1, e_2 and e_3 which satisfy

(1.11) $$\begin{cases} \delta = 3e_3 , \ \varepsilon = (e_1 - e_2)^2 \\ (\delta^2 - \varepsilon)/4 = (e_1 - e_3)(e_2 - e_3) . \end{cases}$$

Continuing, put

$$z = -2x/y , \ w = -2/y$$

as in Silverman [20, Ch. IV], the factor 2 being required since the equation (1.8) in classical Weierstrass form reads $y^2 = 4x^3 + \dots$. Following Silverman [20, Ch. IV], we view $w = w(z)$ as a formal series with coefficients in K satisfying

$$w(z) \equiv z^3 \mod \deg 4 ,$$

and we introduce the standard formal group $F_W(z_1,z_2)$. On the other hand we have the formal group $F_E(X_1,X_2)$ of Euler, which has the explicit form (1.2) and is useful in topological applications. Thus the following result is very desirable.

Theorem 4. *Given elements* δ *and* ε *of a field* K *of characteristic not* 2 *or* 3, *satisfying* $\Delta = \varepsilon(\delta^2 - \varepsilon)^2 \neq 0$, *define* g_2 *and* g_3 *by* (1.10). *Then*

$$X = f(z) = z - \frac{1}{3}\delta w(z) \tag{1.12}$$

is a strict isomorphism from the formal group $F_W(z_1,z_2)$ *of the curve* (1.8) *to the formal group* $F_E(X_1,X_2)$ *of the curve* (1.1). *Moreover, the roots* e_1 , e_2 , e_3 *of the equation* $4x^3 - g_2x - g_3 = 0$ *may be taken to satisfy* (1.11).

This paper is organized as follows. In §2 we deduce Theorem 1 from Theorems 2-4, and also give a simple p-adic proof of the Schur congruences for Legendre polynomials. We prove Theorem 2 in §3, following Gross's argument but with sufficient amplification to make it understandable to anyone with Silverman's book [20] at hand. The complementary Theorem 3 is proved in §4, and then in §5 we prove Theorem 4 by making use of elliptic functions both to suggest the desired transformation and to verify its correctness. The final section gives a more direct proof of the congruences (1.4) and (1.5) of Theorem 1, based on results of Igusa [9, 10] on Jacobi quartics. An elementary verification of Euler's formula for the formal group of a Jacobi quartic is given in the appendix.

My debts are numerous, the foremost being to Bob Stong for an abundance of illuminating correspondence about questions related to this paper, which is an outgrowth of our joint work. I thank David and Gregory Chudnovsky and Dick Gross for answering my questions so elegantly. David Rohrlich gave me much valuable advice, and found the first proof of Theorem 3. I am grateful to Bernard Dwork, Serge Ochanine and Doug Ravenel for providing helpful information. Joe Silverman's beautiful book on elliptic curves [20] has been an indispensable source. Finally, thanks are due to the National Science Foundation and the Institute for Advanced Study for financial support.

§2. *Congruences for Legendre polynomials.* Our main aim, of course, is to deduce the congruences of Theorem 1. We shall begin with some comments on formal groups to explain our interest in such congruences, and then give a quick proof of the Schur congruences, before turning to the main matter.

Let $F(X,Y)$ be a formal group over a commutative ring R , with invariant differential

$$\omega = (1 + \sum_{n\geq 1} w_n T^n)\, dT \,.$$

Fix a prime p , and let u_n be the coefficient of T^{p^n} in

$$[p](T) = pT + \ldots + u_1T^p + \ldots + u_2T^{p^2} + \ldots \,.$$

Moreover, let $v_n (n \geq 1)$ denote the images in R of the Hazewinkel generators v_n [18, A2.2], which by definition satisfy the relations

$$w_{p^n-1} = \sum_{0 \leq i < n} p^{n-1-i} w_{p^i-1} v_{n-i}^{p^i} .$$

Note that in particular

$$v_1 = w_{p-1} , \quad pv_2 = w_{p^2-1} - w_{p-1}^{1+p} ,$$

and that for Euler's formal group over $\mathbf{Z}[\frac{1}{2}][\delta,\varepsilon]$ (and with p odd) we have $w_{2n} = P_n(\delta,\varepsilon)$ and consequently (as in §1)

$$v_1 = P_{(p-1)/2} (\delta,\varepsilon) ,$$

$$v_2 = \frac{1}{p} \{P_{(p^2-1)/2} (\delta,\varepsilon) - P_{(p-1)/2} (\delta,\varepsilon)^{1+p}\} .$$

<u>Lemma 2.1</u>. $u_n \equiv v_n \mod (p,u_1,\ldots,u_{n-1})$.

Note that this justifies the congruences (1.3). In particular, $u_2 = v_2$ for a supersingular elliptic curve over a field of characteristic p .

<u>Proof</u>. We refer to the appendix on formal groups in Ravenel's book [18, A2]. As noted in Theorem A2.2.3 there, the generators of Hazewinkel and Araki agree mod p . Hence it suffices to deal with Araki's generators in the universal case, so that we can take advantage of the formula A2.2.4:

$$(2.1) \qquad [p](T) = pT +_F v_1 T^p +_F \ldots +_F v_n T^{p^n} +_F \ldots .$$

Since u_n is the coefficient of T^{p^n} in $[p](T)$, we want to find the coefficient of T^{p^n} in the right hand side of (2.1) modulo the ideal $(p,u_1,\ldots,u_{n-1})$, which by induction coincides with the ideal $(p,v_1,\ldots,v_{n-1})$. Evidently this coefficient is v_n , since

$$pT +_F v_1 T^p +_F \ldots +_F v_{n-1} T^{p^{n-1}} \equiv 0$$

mod $(p,v_1,\ldots,v_{n-1})$. □

We turn next to the <u>Schur congruences</u>, which assert that if p is an odd prime and

$$n = n_0 + n_1 p + \ldots + n_r p^r$$

with $0 \leq n_i < p$ is the p-adic expansion of a positive integer n , then the Legendre polynomials satisfy

$$(2.2) \qquad P_n(x) \equiv P_{n_0}(x)\, P_{n_1}(x)^p \ldots P_{n_r}(x)^{p^r} \pmod p .$$

Despite their name, the first published proof of these congruences was apparently given by Wahab [21]; they appear in a 1924 Ph.D. thesis by H. Ille, a student of Schur.

We shall present a brief proof in order to show that the Schur congruences may be viewed as a simple p-adic phenomenon.

Let p be an odd prime, and view the power series ring $\mathbb{F}_p[[x]]$ as complete and Hausdorff in the usual way, i.e. in the (x)-adic topology, so that one can take limits in $\mathbb{F}_p[[x]]$.

<u>Lemma 2.2</u>. In $\mathbb{F}_p[[x]]$, <u>one has</u> $(1+x)^{-\frac{1}{2}} = \lim_{\ell\to\infty} [(1+x)^{(p-1)/2}]^{1+p+\ldots+p^{\ell-1}}$.

<u>Proof</u>. Equivalently, we are to show that

$$\lim_{\ell\to\infty} [(1+x)^{\frac{1}{2}}]^{(p-1)(1+p+\ldots+p^{\ell-1}) + 1} = 1 ,$$

i.e. that

$$\lim_{\ell\to\infty} [(1+x)^{\frac{1}{2}}]^{p^{\ell}} = 1 ,$$

which is clear since $(1+x)^{p^{\ell}} = 1 + x^{p^{\ell}}$ in $\mathbb{F}_p[[x]]$. □

Putting $R(t) = 1 - 2xt^2 + t^4$, we obtain at once

$$R(t)^{-\frac{1}{2}} \equiv \lim_{\ell\to\infty} [R(t)^{(p-1)/2}]^{1+p+\ldots+p^{\ell-1}} \pmod p , \tag{2.3}$$

which first yields

$$\sum_{n=0}^{p-1} P_n(x)\, t^{2n} \equiv (1 - 2\delta t^2 + t^4)^{(p-1)/2} \pmod p \tag{2.4}$$

and then

$$\sum_{n\geq 0} P_n(x)\, t^{2n} \equiv \lim_{\ell\to\infty} \Big(\sum_{n=0}^{p-1} P_n(x)\, t^{2n}\Big)^{1+p+\ldots+p^{\ell-1}} \tag{2.5}$$

(mod p) . The Schur congruences are immediate, by a comparison of the coefficients of t^{2n} in (2.5), while (2.4) yields the further well-known congruences

$$P_n(x) \equiv P_{p-1-n}(x) \pmod p$$

for $0 \leq n < p$.

Note that the Schur congruences imply that the polynomial $V_2(x)$ introduced in (1.3') lies in $\mathbb{Z}[\frac{1}{2}][x]$. In fact, this is already a consequence of the integrality of the Hazewinkel generators.

<u>We now turn to the proof of Theorem 1</u>. The congruences (1.4) and (1.5) are evident when $p = 3$: one then has $V_1(x) = x$ and $V_2(x) = \frac{1}{8}(9x^4 - 10x^2 + 1)$. We now <u>assume $p > 3$</u> and establish the congruences (1.4) and (1.5').

The following result implies that $\mathbb{F}_p[x]/(V_1(x))$ is a direct sum of finite fields, so that the congruences (1.4) and (1.5') will follow from the corresponding equalities for supersingular Jacobi quartics. I thank Bernard Dwork for pointing out this well-known argument due to Igusa.

<u>Lemma 2.3</u>. <u>For an odd prime</u> p <u>and</u> $0 < n < p$, <u>the Legendre polynomial</u> $P_n(x)$

<u>has no repeated roots</u> mod p .

<u>Proof</u>. One sees from the generating function that $P_n(\pm 1) = (\pm 1)^n$. Since $P_n(x)$ satisfies Legendre equation ([11])

$$(1 - x^2)\, P_n''(x) - 2xP_n'(x) + n(n+1)\, P_n(x) = 0 ,$$

if x_0 were a repeated root of $P_n(x)$, then $0 = P_n(x) = P_n'(x)$ would imply all derivatives vanish at x_0 (note that $x_0 \neq \pm 1$) , hence all coefficients would vanish since $n < p$, a contradiction to $P_n(1) = 1$. □

Thus it will suffice to show that for a <u>supersingular</u> Jacobi quartic (with $\varepsilon = 1$)

$$Y^2 = 1 - 2\delta X^2 + X^4 \tag{2.6}$$

over a field K of characteristic $p > 3$, so that

$$\Delta = (\delta^2 - 1)^2 \neq 0$$

while

$$v_1 = u_1 = P_{(p-1)/2}(\delta) = V_1(\delta) = 0 ,$$

we have equalities

$$V_2(\delta) = \left(\frac{-1}{p}\right) \quad \text{and} \tag{2.7}$$

$$(\delta^2 - 1)^{(p^2-1)/12} = 1 . \tag{2.8}$$

We now apply Theorem 4, and suggest the reader review its statement given in the introduction. Since we have taken $\varepsilon = 1$, our "dictionary" (1.11) reads

$$\begin{cases} \delta = 3e_3 , \quad (e_1 - e_2)^2 = 1 \\ \delta^2 - 1 = 4(e_1 - e_3)(e_2 - e_3) \end{cases} \tag{2.9}$$

Theorems 2 and 3 provide the equalities

$$u_2 = \left(\frac{-1}{p}\right) \Delta^{(p^2-1)/12} \tag{2.10}$$

and

$$\Delta^{(p^2-1)/24} = (e_1 - e_2)^{(p^2-1)/4} . \tag{2.11}$$

Since $\Delta = (\delta^2 - 1)^2$ when $\varepsilon = 1$, we see that

$$u_2 = \left(\frac{-1}{p}\right) (\delta^2 - 1)^{(p^2-1)/6} \tag{2.12}$$

and

$$(\delta^2 - 1)^{(p^2-1)/12} = 1 . \tag{2.13}$$

In particular, $(\delta^2 - 1)^{(p^2-1)/6} = 1$, and so

(2.14) $$u_2 = \left(\frac{-1}{p}\right) .$$

Since the formal groups $F_E(X_1,X_2)$ and $F_W(z_1,z_2)$ are strictly isomorphic and have height 2 ($u_1 = 0$ for both), the elements u_2 for both coincide. Thus $u_2 = v_2 = V_2(\delta)$, and so we obtain

(2.15) $$V_2(\delta) = \left(\frac{-1}{p}\right) .$$

Having verified (2.7) and (2.8), the deduction of Theorem 1 from Theorems 2, 3 and 4 is complete. □

§3. The formal group of a supersingular elliptic curve. We turn to the proof of Theorem 2, and so consider an elliptic curve with Weierstrass equation

$$y^2 + a_1xy + a_3y = x^3 + a_2x^2 + a_4x + a_6 .$$

We also draw attention to the coefficients u_1 and u_2 in the expansion

$$[p](z) = pz + \ldots + u_1z^p + \ldots + u_2z^{p^2} + \ldots ,$$

p a prime.

The cases $p = 2$ and $p = 3$ can be read off from [20, Chapters III, IV], as congruences mod (p,u_1) in $\mathbb{Z}[a_1,\ldots,a_6]$. For $p = 2$, one has $u_1 \equiv a_1 \pmod 2$ and then $u_2 \equiv a_3 \bmod (2,u_1)$, while $\Delta \equiv a_3^4 \bmod (2,u_1)$. For $p = 3$, we can assume $a_1 = a_3 = 0$, and then $u_1 \equiv a_2 \pmod 3$, $u_2 \equiv -a_4^2 \bmod (3,u_1)$ and $\Delta \equiv -a_4^3 \bmod (3,u_1)$.

Taking $p > 3$, there are two special cases $j = 0$, $j = 1728$ requiring separate treatment. The elliptic curve with equation

$$y^2 = x^3 + x$$

has $\Delta = -64$ and $j = 1728$ and invariant differential

$$\omega = (1 - 4z^4)^{-\frac{1}{2}} dz$$
$$= \sum_{n \geq 0} \binom{2n}{n} z^{4n} dz .$$

In general, if an elliptic curve has invariant differential

$$\omega = (1 + \sum_{n \geq 1} w_n z^n) dz$$

then (see §2) $u_1 \equiv w_{p-1} \pmod p$ and $u_2 \equiv \frac{1}{p}(w_{p^2-1} - w_{p-1}^{1+p}) \bmod (p,u_1)$. We see that the elliptic curve in question is supersingular if and only if $p \equiv 3 \pmod 4$, and that in this case one is being asked to verify that

$$\binom{(p^2-1)/2}{(p^2-1)/4} + p \equiv 0 \pmod{p^2} .$$

Similarly, the elliptic curve with equation

$$y^2 + y = x^3$$

has $\Delta = -27$, $j = 0$ and invariant differential

$$\omega = (1 - 4z^3)^{-\frac{1}{2}}\, dz$$
$$= \sum_{n \geq 0} \binom{2n}{n} z^{3n} .$$

We see that the elliptic curve in question is supersingular if and only if $p \equiv 2$ mod 3 , and an easy calculation involving quadratic reciprocity shows that one is being asked to verify that

$$\binom{2(p^2-1)/3}{(p^2-1)/3} + p \equiv 0 \pmod{p^2} .$$

We see that the special cases $j = 0$, 1728 follow from

<u>Proposition 3.1. If an odd prime</u> p <u>satisfies</u> $p \equiv -1 \mod n$, <u>then one has</u>

$$\binom{2(p^2-1)/n}{(p^2-1)/n} + p \equiv 0 \pmod{p^2} .$$

<u>Proof</u>. If $m = a_0 + a_1 p + \ldots + a_s p^s$, $0 \leq a_i < p$, then the power of p dividing $m!$ is

$$\nu = \nu_p(m!) = (m - (a_0 + \ldots + a_s))/(p - 1) ,$$

and one has

$$m! \equiv (-p)^{\nu}\, a_0!\, a_1! \ldots a_s! \pmod{p^{\nu+1}} .$$

If $p \equiv -1 \mod n$ then $p + 1 = nk$ gives

$$(p^2 - 1)/n = (p - k) + (k - 1)\, p \; ; \; \nu = k - 1$$
$$2(p^2 - 1)/n = (p - 2k) + (2k - 1)\, p \; ; \; \nu = 2k - 1$$

and then mod p^2 one has congruences

$$\binom{2(p^2-1)/n}{(p^2-1)/n} \equiv (-p) \frac{(2k-1)!\ (p-2k)!}{(p-k)!^2\ (k-1)!^2}$$
$$\equiv (-p)\binom{p-1}{k-1}^2 / \{\binom{p-1}{2k-1}(p-1)!\}$$
$$\equiv (-p)(-1)^{2(k-1)}/\{(-1)^{2k-1}\ (-1)\}$$
$$\equiv -p . \qquad \square$$

<u>Note</u>: This argument is due to Serge Ochanine, with the congruence for $m!$ going back to Stickelberger and Hensel (1890 and 1902) or Anton (1869) for $s = 1$.

Having disposed of special cases, we assume that $p > 3$ and that $j \neq 0$, 1728.

We shall follow Gross's proof and provide full details. Since we are dealing with a supersingular elliptic curve E , we have

$$[p](z) = u_2 z^{p^2} + \ldots$$

with $u_2 \neq 0$, and the j-invariant $j(E)$ lies in $\mathbb{F}_{p^2}$ ([20, Thm . V. 3.1]). Applying [20, Prop. III. 1.4], we see that E is isomorphic to an elliptic curve defined over $\mathbb{F}_{p^2}$. Since $u_2/\Delta^{(p^2-1)/12}$ is an isomorphism invariant for supersingular elliptic curves, we can assume from the start that E is defined over $\mathbb{F}_{p^2}$.

As a final preliminary observation, note that an isogeny $\phi\colon E \to E'$ of elliptic curves induces a homomorphism $f\colon F \to F'$ of formal groups, in a natural way; i.e., one has

$$f(F(X,Y)) = F'(f(X),f(Y))$$

as formal series. As special cases, for the multiplication-by-m map $[m]\colon E \to E$ it is evident that the corresponding homomorphism of formal groups is $f(T) = [m](T)$. Also, for a Frobenius morphism $\phi_q\colon E \to E^{(q)}$ ([20, III. 4.6]) it is evident that $f(T) = T^q$.

We now examine the morphism $[p]\colon E \to E$, E being supersingular and defined over $\mathbb{F}_{p^2}$. By [20, Thm . IV. 7.4], $[p]$ has inseparable degree p^2 , and then [20, Cor. II. 2.12] provides a factorization

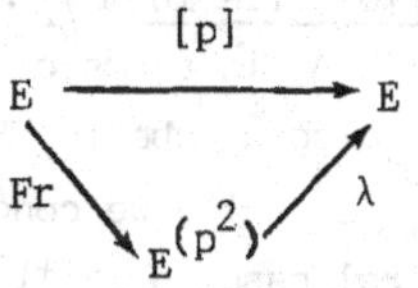

$$
\begin{array}{ccc}
& [p] & \\
E & \longrightarrow & E \\
& \searrow \quad \nearrow & \\
\text{Fr} & E^{(p^2)} & \lambda
\end{array}
$$

where Fr denotes the p^2-power Frobenius and λ is a separable morphism. Since E is defined over $\mathbb{F}_{p^2}$, $E^{(p^2)} = E$ and Fr is an endomorphism of degree p^2 ([20, Prop. II. 2.11]). Since also $[p]$ has degree p^2 ([20, Prop. III. 6.4]) and degrees multiply, $\deg(\lambda) = 1$ and so λ is an automorphism of E . Since we're assuming that $j(E) \neq 0$, 1728 , [20, Thm . III. 10.1] implies that $\lambda = [\pm 1]$. In particular, we have learned that

$$\text{Fr} = [\pm p]\colon E \to E ,$$

which we can rewrite as

$$[p] = [\pm 1] \circ \text{Fr} .$$

Passing to homomorphisms of formal groups (see the previous paragraph), we find that

$$u_2 T^{p^2} + \text{higher terms} = [\pm 1](T^{p^2}) ,$$

and so $u_2 = \pm 1$. Hence we have

$$\text{Fr} = [u_2 p]\colon E \to E .$$

We next make the assumption that

$$u_2 = \left(\frac{-1}{p}\right) ,$$

a hypotheses which will be removed at the end of the argument (by twisting). Now our aim is to show that $\Delta^{(p^2-1)/12} = 1$. Our assumption implies that

$$\mathrm{Fr} = \left[\left(\frac{-1}{p}\right) p\right] : E \to E ,$$

and we observe that

$$\left(\frac{-1}{p}\right) p \equiv 1 \pmod 4 .$$

Hence Fr acts trivially on $E[4]$, the 4-torsion subgroup of E . As noted in [20, III. 4.6], the set of points fixed by Fr is exactly $E(\mathbb{F}_{p^2})$. Hence <u>the coordinates of all points in $E[4]$ lie in $\mathbb{F}_{p^2}$</u> .

We now refer to Lang and Trotter [15, pp. 218-220] for what must be a very classical formula giving a root $\Delta^{\frac{1}{4}}$ of Δ in the field $K(E[4])$ of "4-division points," valid provided K has characteristic different from 2 and 3. In view of the conclusion above, $\Delta \in \mathbb{F}_{p^2}$ is a fourth power in $\mathbb{F}_{p^2}$, and so $\Delta^{(p^2-1)/4} = 1$.

Now consider the action of Fr on $E[3]$, the 3-torsion points. Since $p > 3$ and $\mathrm{Fr} = [\pm p]$, and since the negative $-P$ of each $P = (x,y) \in E$ has the same x-coordinate, we see that Fr fixes the x-coordinates of all points in $E[3]$. Thus <u>the x-coordinates of all points in $E[3]$ lie in $\mathbb{F}_{p^2}$</u>. We now refer to Serre [19, p. 305] for an explicit root $\Delta^{1/3}$ of Δ in terms of x-coordinates of points in $E[3]$. Thus $\Delta^{1/3} \in \mathbb{F}_{p^2}$, and Δ is also a cube in $\mathbb{F}_{p^2}$, i.e. $\Delta^{(p^2-1)/3} = 1$. Since $d = \Delta^{(p^2-1)/12} \in \mathbb{F}_{p^2}$ satisfies $d^3 = d^4 = 1$, we conclude that $\Delta^{(p^2-1)/12} = 1$.

It remains to deal with the general case $u_2 = \pm 1$. For an elliptic curve

$$E:\ y^2 = x^3 + Ax + B$$

and $D \in K^*$, one forms the twisted curve

$$E_D:\ y^2 = x^3 + D^2Ax + D^3B .$$

The discriminants and j-invariants of these curves satisfy

$$\Delta_D = D^6\Delta , \quad j_D = j .$$

In particular, $(\Delta_D)^{(p^2-1)/12} = -\Delta^{(p^2-1)/12}$ for half the values of $D \in \mathbb{F}_{p^2}^*$, namely when D is a non-square. Moreover, since u_2 is a polynomial of weight $p^2 - 1$ in the coefficients of the Weierstrass equation, it is clear that for E_D one has

$$(u_2)_D = D^{(p^2-1)/2} u_2 ;$$

in particular, $(u_2)_D = -u_2$ if D is a non-square in $\mathbb{F}_{p^2}^*$. It is now clear that the above conclusions under the hypothesis $u_2 = \left(\frac{-1}{p}\right)$ imply the result for $u_2 = \pm 1$.

This completes the proof of Theorem 2, and explains the appearance of the factor $\left(\frac{-1}{p}\right)$. □

§4. <u>A formula for</u> $\Delta^{(p^2-1)/24}$ <u>in the supersingular case</u>. We next turn to the proof of Theorem 3, and so take a supersingular elliptic curve

$$
\begin{aligned}
y^2 &= 4x^3 - g_2x - g_3 \\
&= 4(x - e_1)(x - e_2)(x - e_3)
\end{aligned}
\tag{4.1}
$$

over a field of characteristic $p > 3$. <u>We shall assume that</u> $u_2 = \left(\frac{-1}{p}\right)$ <u>in the following</u>, hence $\Delta^{(p^2-1)/12} = 1$ by Theorem 2; the general case follows by twisting, as at the end of the previous section. We shall observe that

$$(e_\alpha - e_\beta)^{(p^2-1)/2} = 1 \quad (\alpha \neq \beta) , \tag{4.2}$$

and go on to argue that

$$\Delta^{(p^2-1)/24} = (e_\alpha - e_\beta)^{(p^2-1)/4} \quad (\alpha \neq \beta) . \tag{4.3}$$

As to (4.2), recall that $u_2 = \left(\frac{-1}{p}\right)$ implies that the coordinates of all points in $E[4]$ (and so also $E[2]$) lie in $\mathbb{F}_{p^2}$. Hence each e_α lies in $\mathbb{F}_{p^2}$; since $(p^2-1)/2$ is even, it will suffice to argue that $(e_i - e_{i-1})^{(p^2-1)/2} = 1$, using cyclic indices mod 3 as in Lang and Trotter [15, pp. 218-220]. Indeed, Lang and Trotter write down a square root w_i of $e_i - e_{i-1}$ which is expressed rationally in terms of the coordinates in $E[4]$, and so lies in $\mathbb{F}_{p^2}$, whence $(e_i - e_{i-1})^{(p^2-1)/2} = 1$ as required.

For the verification of (4.3), we shall need to consider the 8-torsion points on E. Since the curve E is supersingular and we assume $u_2 = \left(\frac{-1}{p}\right)$, we observe as in §3 that

$$Fr = \left[\left(\frac{-1}{p}\right) p\right] : \quad E \to E$$

with $\left(\frac{-1}{p}\right) p \equiv 1 \pmod 4$. Hence Fr acts as either the identity or as $[5]$ on $E[8]$. The argument must take both possibilities into account, but will not be broken into cases.

We introduce notation as in [15, pp. 218-220], recalling first that the points of order 2 on E are given in terms of coordinates by

$$(e_i, 0) \quad , \quad i = 1,2,3 .$$

The points P of order 4 such that $2P = (e_i, 0)$ have the form

$$
\begin{aligned}
&(e_i + u_i \, , \pm v_i) \ \text{ where } \ v_i^2 = u_i^2(3e_i + 2u_i) , \\
&(e_i - u_i \, , \pm v_i') \ \text{ where } \ v_i'^2 = u_i^2(3e_i - 2u_i) ;
\end{aligned}
$$

here u_i satisfies $u_i^2 = 3e_i^2 - \frac{1}{4} g_2$. Then define $w_i = \dfrac{v_i - v_i'}{2u_i}$ and $W = w_1w_2w_3$, observing that

$$w_i^2 = e_i - e_{i-1} \quad , \quad 16W^4 = \Delta ;$$

hence $\Delta^{\frac{1}{4}} = 2W \in K(E[4])$, as was used in §3.

<u>Lemma 4.1. The formulas</u> (4.3) <u>follow from the assertion that each</u> u_i <u>is a square in</u> $\mathbb{F}_{p^2}$.

<u>Proof</u>. We compute:

$$\Delta^{(p^2-1)/24} = [(2W)^4]^{(p^2-1)/24} = (2W)^{(p^2-1)/6}$$

and know that this is ± 1 . Since $(\pm 1)^3 = \pm 1$, we have

$$\begin{aligned}\Delta^{(p^2-1)/24} &= [(2W)^{(p^2-1)/6}]^3 = (2W)^{(p^2-1)/2} \\ &= 2^{(p^2-1)/2} (w_1 w_2 w_3)^{(p^2-1)/2} \\ &= (w_1 w_2 w_3)^{(p^2-1)/2}\end{aligned}$$

since

$$2^{(p^2-1)/2} = (2^{p-1})^{(p+1)/2} = 1$$

in $\mathbb{F}_p$.

Observe next that

$$\begin{aligned}u_i^2 &= 3e_i^2 - \tfrac{1}{4} g_2 \\ &= (e_i - e_{i-1})(e_i - e_{i+1}) \\ &= (-1)\, w_i^2 w_{i+1}^2 ,\end{aligned}$$

so that

$$u_i = \sqrt{-1}\, w_i w_{i+1} .$$

Noting that $(\sqrt{-1})^{(p^2-1)/2} = 1$, so that $\sqrt{-1}$ has a square root in $\mathbb{F}_{p^2}$, the hypothesis that u_i is a square in $\mathbb{F}_{p^2}$ implies that the same is true of $w_i w_{i+1}$. Therefore

$$w_1^{(p^2-1)/2} = w_2^{(p^2-1)/2} = w_3^{(p^2-1)/2} ,$$

and so we may return to the previous paragraph and conclude that

$$\Delta^{(p^2-1)/24} = [(w_i)^{(p^2-1)/2}]^3 = w_i^{(p^2-1)/2}$$

(using $(\pm 1)^3 = \pm 1$ again), i.e.

$$\Delta^{(p^2-1)/24} = (e_i - e_{i-1})^{(p^2-1)/4} ,$$

which is equivalent to (4.3) since $(p^2-1)/4$ is even. □

It remains to argue that u_i is a square in $\mathbb{F}_{p^2}$; the following argument for u_1 applies also to u_2 and u_3 . Let $P = (x,y)$ have order 8 and satisfy $[2]P = (e_1 + u_1, v_1)$; then write

$$[5]P = P + (e_1,0) = (x',y') .$$

We shall argue that

$$\tag{4.4} \tfrac{1}{2}(x + x') = e_1 + u_1 \pm u_1^{\frac{1}{2}} \{(e_1 - e_2)^{\frac{1}{2}} + (e_1 - e_3)^{\frac{1}{2}}\} .$$

This formula expresses $u_1^{\frac{1}{2}}$ rationally in terms of the coordinates of elements of $E[8]$, and in such a way that <u>the expression is invariant under</u> $[5]$ <u>acting on</u> $E[8]$. We conclude that $u_1^{\frac{1}{2}}$ lies in $\mathbb{F}_{p^2}$; the same holds for $u_2^{\frac{1}{2}}$ and $u_3^{\frac{1}{2}}$, and then Lemma 4.1 yields (4.3).

Turning to the verification of (4.4), one sees easily by means of the formula

$$\{(e_1 - e_2)^{\frac{1}{2}} + (e_1 - e_3)^{\frac{1}{2}}\}^2 = 2u_1 + 3e_1$$

that it will suffice to verify the formula

$$\tag{4.5} (x + x')^2 - 4(x + x')(e_1 + u_1) + 4(e_1 + u_1)^2 = u_1(2u_1 + 3e_1) .$$

In fact, it is easy to express $x + x'$ in terms of x, namely one has

$$\tag{4.6} x' = e_1 + \frac{(e_1 - e_2)(e_1 - e_3)}{x - e_1}$$

which at once yields

$$\tag{4.7} x + x' = \frac{x^2 + e_1^2 + e_2e_3}{x - e_1} .$$

The verification of (4.6) can be found in Whittaker and Watson [22, §20.33], by straight algebra from the addition theorem. Putting (4.7) into (4.5) and clearing fractions yields a quartic equation for x; on the other hand, there are four points $P = (x,y)$ satisfying $[2]P = (e_1 + u_1, v_1)$, and they have distinct x-coordinates which are the roots of a second quartic equation that one writes down by means of the duplication formula [20, III. 2.3d]. A patient comparison shows that these quartic equations are identical, and so we have verified (4.4). As at the end of §3, we can remove the hypothesis $u_2 = \left(\frac{-1}{p}\right)$ by twisting, and have completed the proof of Theorem 3. □

<u>Note</u>: One can also use transcendental methods to verify (4.4), by showing (with guidance from [1] or Copson [6]) that for a half-period ω_1 of $\wp(u)$ with

$$\wp(\omega_1) = e_1 \quad \text{and} \quad \wp(\omega_1/2) = e_1 + u_1$$

one has

$$\tfrac{1}{2}\left\{\wp\left(\frac{\omega_1}{4}\right) + \wp\left(\frac{5\omega_1}{4}\right)\right\} = e_1 + u_1 + u_1^{\frac{1}{2}} \{(e_1 - e_2)^{\frac{1}{2}} + (e_1 - e_3)^{\frac{1}{2}}\} .$$

§5. <u>The transformation to quartic form</u>. Our aim in this section is to prove Theorem 4. We shall make use of elliptic functions both to suggest the desired transformation and to verify its correctness.

Begin in the classical setting, with $\delta, \varepsilon \in \mathbb{C}$ for which $\Delta = \varepsilon(\delta^2 - \varepsilon)^2 \neq 0$. Then one has (see Ochanine [17]) an odd elliptic function $s(u)$ of order 2 with period lattice 2Ω and for which

$$(X,Y) = (s(u), s'(u))$$

satisfies

$$Y^2 = 1 - 2\delta X^2 + \varepsilon X^4 , \tag{5.1}$$

i.e. $s(u)$ satisfies the corresponding differential equation. Moreover, Ω is generated by two poles ω_1, ω_2 of $s(u)$; putting $\omega_3 = \omega_1 + \omega_2$, $s(u)$ has simple zeros at $u = 0$ and $u = \omega_3$. Finally, $s'(0) = 1$ and $s(u + \omega_3) = -s(u)$.

Now let $P(u)$ be the Weierstrass function for the lattice 2Ω, and put $e_i = P(\omega_i)$, $i = 1,2,3$ as is customary. One sees at once that

$$s(u) = -2(P(u) - e_3)/P'(u) . \tag{5.2}$$

Indeed, $s(u)\,P'(u)$ and $P(u) - e_3$ are both even elliptic functions with double poles at the origin and zeros of order 2 at ω_3 (and no other zeros or poles modulo 2Ω), hence are multiples of one another, the factor being -2 as one sees by expanding about the origin. As usual,

$$(x,y) = (P(u), P'(u))$$

satisfies

$$y^2 = 4x^3 - g_2 x - g_3 \tag{5.3}$$

for appropriate constants g_2 and g_3. Indeed a straightforward comparison of coefficients, for which

$$s(u) = u - \frac{\delta}{3} u^3 + \frac{\delta^2 + 3\varepsilon}{30} u^5 + \ldots$$

and

$$P(u) = u^{-2} + \frac{g_2}{20} u^2 + \frac{g_3}{28} u^4 + \ldots$$

are sufficient, leads to the formulas (cf. (1.10) and (1.11) in the introduction)

$$g_2 = \frac{1}{3}(\delta^2 + 3\varepsilon) , \quad g_3 = \frac{\delta}{27}(\delta^2 - 9\varepsilon) \tag{5.4}$$

and

$$\left\{ \begin{array}{l} \delta = 3e_3 , \quad \varepsilon = (e_1 - e_2)^2 \\ (\delta^2 - \varepsilon)/4 = (e_1 - e_3)(e_2 - e_3) \end{array} \right. \tag{5.5}$$

so that in particular the discriminant is given by

$$\Delta = \varepsilon(\delta^2 - \varepsilon)^2 = 16(e_1 - e_2)^2 (e_1 - e_3)^2 (e_2 - e_3)^2 .$$

Note also that $u = 0$ (the origin on $\mathbb{C}/2\Omega$) corresponds to the origin at ∞ on the curve (5.3), and to the origin $(X,Y) = (0,1)$ on the Jacobi quartic (5.1).

We now turn to the algebraic setting of Theorem 4, and so begin with a non-singular Jacobi quartic (5.1) over a field K of characteristic not 2 or 3. Define g_2 and g_3 by means of (5.4), so we have also the elliptic curve given by the Weierstrass equation (5.3). One now easily verifies the further formulas (5.5), which distinguish one of the roots of the equation $4x^3 - g_2x - g_3 = 0$.

With our attention now on the formal groups, a slight modification of [20, Ch. IV] leads us to introduce the variables

$$z = -2x/y \ , \ w = -2/y \tag{5.6}$$

in place of (x,y) , so that

$$x = z/w \ , \ y = -2/w \tag{5.6'}$$

and the origin on the curve (5.3) becomes $(z,w) = (0,0)$. Moreover, one views $w = w(z) \in K[[z]]$ as a formal series satisfying

$$w(z) \equiv z^3 \mod \deg 4 \ .$$

Observe now that (5.2) and (5.6) suggest the transformation

$$\begin{aligned} X = f(z) &:= z - e_3w(z) \\ &= z - \frac{1}{3}\,\delta w(z) \ . \end{aligned} \tag{5.7}$$

Indeed, we shall proceed to show that $f(z)$ is the desired strict isomorphism from the formal group $F_W(z_1,z_2)$ of the curve (5.3), with uniformizing parameter z at the origin, to the formal group $F_E(X_1,X_2)$ of the curve (4.1), for which X serves as uniformizing parameter at the origin. Thus we wish to establish the identity of formal series

$$f(F_W(z_1,z_2)) = F_E(f(z_1),f(z_2)) \tag{5.8}$$

over the field K .

One sees easily, using (5.4), that all coefficients appearing in (5.8) have expressions as polynomials in $\mathbb{Z}[1/6][\delta,\varepsilon]$. By verifying (5.8) whenever δ and ε are complex numbers with $\varepsilon(\delta^2 - \varepsilon)^2 \neq 0$, it will follow that the corresponding polynomials in $\mathbb{Z}[1/6][\delta,\varepsilon]$ are identical, so that (5.8) holds over any field K of characteristic different from 2 and 3.

It remains to verify (5.8) with $\delta,\varepsilon \in \mathbb{C}$ and $\varepsilon(\delta^2 - \varepsilon)^2 \neq 0$. Guided by the formulas (5.6), introduce elliptic functions

$$z(u) = -2\wp(u)/\wp'(u) \ , \ w(u) = -2/\wp'(u) \ . \tag{5.9}$$

Note that near $u = 0$ we have

$$z(u) = u + \ldots \ , \ w(u) = u^3 + \ldots$$

Moreover (5.2) yields

$$s(u) = z(u) - e_3w(u) = f(z(u)) \ . \tag{5.10}$$

Note next that the formal groups yield classical addition formulas

(5.11) $$s(u_1 + u_2) = F_E(s(u_1), s(u_2))$$

and

(5.12) $$z(u_1 + u_2) = F_W(z(u_1), z(u_2)) .$$

The last three formulas imply that

$$f(F_W(z(u_1), z(u_2))) = F_E(f(z(u_1)), f(z(u_2)))$$

for u_1 and u_2 near zero. Since $z(u) = u + \dots$ and so $z(u)$ has a holomorphic inverse near $z = 0$, we conclude that (5.8) holds as an identity between holomorphic functions for z_1, z_2 near zero, and so as a formal series. We have now completed the proof of Theorem 4. □

§6. A direct proof of Theorem 1. The methods of Igusa [9, 10] for dealing with Jacobi quartics

(6.1) $$Y^2 = 1 - 2\delta X^2 + X^4$$

permit one to prove the following congruences. As above, let p be an odd prime and write

$$[p](X) = pX + \dots + u_1 X^p + \dots + u_2 X^{p^2} + \dots$$

for the multiplication-by-p series of the corresponding formal group. As in §2, we have

$$u_1 \equiv P_{(p-1)/2}(\delta) \bmod p .$$

The congruences in question are

(6.2) $$u_2 \equiv \left(\frac{-1}{p}\right) \quad \bmod (p, u_1)$$

and

(6.3) $$(\delta^2 - 1)^{(p^2-1)/4} \equiv 1 \bmod (p, u_1) .$$

As noted previously, the discriminant for the Jacobi quartic is $\Delta = (\delta^2 - 1)^2$, taking $\varepsilon = 1$. Hence (6.3) states that

$$\Delta^{(p^2-1)/8} \equiv 1 \bmod (p, u_1) .$$

Arguing as at the end of §3, we conclude that for $p > 3$ we also have

$$\Delta^{(p^2-1)/3} \equiv 1 \bmod (p, u_1) ,$$

and so in fact

(6.4) $$\Delta^{(p^2-1)/24} \equiv 1 \bmod (p, u_1) ;$$

i.e. for $p > 3$ we have

(6.4') $$(\delta^2 - 1)^{(p^2-1)/12} \equiv 1 \mod (p, u_1) .$$

In view of the preliminary results in §2, these congruences are equivalent to those of Theorem 1.

We begin with the proof of (6.2) based on Igusa [10, §1], in which one finds a convenient summary of certain results of his earlier paper [9]. For an odd integer n, one has

$$[n](X) = (-1)^{(n-1)/2} \, T_n(X)/F_n(X)$$

for polynomials $F_n(X)$ and $T_n(X)$ with coefficients in $\mathbf{Z}[\delta]$, where

$$T_n(X) = X^{n^2} \, F_n(X^{-1}) .$$

When n is an odd prime p, one writes $P(\delta) = P_{(p-1)/2}(\delta)$ and finds that mod p

$$T_p(X) \equiv X^p \, \{(X^p)^{p-1} + \sum_{0<2i<p-1} P(\delta)\gamma_i(\delta)(X^p)^{2i} + \left(\frac{-1}{p}\right) P(\delta)\}$$

with $\gamma_i \in \mathbf{Z}[\delta]$. From this it follows that mod p

$$F_p(X) \equiv 1 + \sum_{0<2i<p-1} P(\delta)\gamma_i(\delta)(X^p)^{p-1-2i} + \left(\frac{-1}{p}\right) P(\delta)X^{p(p-1)} .$$

Going back to

$$[p](X) = \left(\frac{-1}{p}\right) T_p(X)/F_p(X)$$
$$\equiv P(\delta)X^p + \text{higher terms}$$

mod p, it is clear from the expressions above that

(6.5) $$[p](X) \equiv \left(\frac{-1}{p}\right) X^{p^2} \mod (p, P(\delta)) .$$

This yields (6.2), which is equivalent to the first congruence of Theorem 1 by Lemma 2.1. In addition, if u_n denotes the coefficient of X^{p^n} in $[p](X)$ then also

(6.6) $$u_n \equiv 0 \mod (p, u_1)$$

for $n > 2$.

We now turn to the proof of (6.3). By Lemma 2.3, it suffices to establish the equality

(6.3') $$(\delta^2 - 1)^{(p^2-1)/4} = 1$$

for a <u>supersingular</u> Jacobi quartic over a field of characteristic $p > 2$, i.e. one for which $P(\delta) = 0$. Note that the analog of (6.3') for a Jacobi quartic

(6.1') $$Y^2 = 1 - 2\delta X^2 + \varepsilon X^4$$

is

(6.3") $$(\delta^2 - \varepsilon)^{(p^2-1)/4} = \varepsilon^{(p^2-1)/4} ,$$

an equality of modular forms of weight $p^2 - 1$ and level 2. Thus the validity of (6.3") for one Jacobi quartic implies its truth for all isomorphic Jacobi quartics. Taking $\varepsilon = 1$ again, we shall now begin the proof of (6.3').

Following Igusa [9, §2], let E be an elliptic curve with corresponding supersingular Jacobi quartic (6.1). Then E is supersingular, whence as in §3 one has $j(E) \in \mathbb{F}_{p^2}$ and so we can assume that E is defined over $\mathbb{F}_{p^2}$. In order to verify (6.3'), it will suffice to show that $\delta^2 - 1$ lies in $\mathbb{F}_{p^2}$ and moreover has a fourth root in $\mathbb{F}_{p^2}$.

We therefore undertake some elementary algebra to locate a square root and then a fourth root of $\delta^2 - 1$. Following [9, §2], one chooses points r and s of order 4 on E, with $2r \neq 2s$, and introduces a rational function x on E with divisor

$$\mathrm{div}(x) = 2*(0) - 2*(2r) .$$

Then x is an odd function on E, unique up to sign if one also requires that

(6.7) $$x(u + r)\, x(u) = 1 \quad (u \in E) .$$

Moreover, x is invariant under translation by 2r, but changes sign under translation by 2s. Then

$$\delta = \tfrac{1}{2}(\sigma^2 + \sigma^{-2}) ,$$

where $\sigma = x(s)$. It follows at once that

(6.8) $$\delta^2 - 1 = \left(\frac{\sigma^2 - \sigma^{-2}}{2}\right)^2 .$$

Thus we shall want to know that $\sigma^2 \in \mathbb{F}_{p^2}$. Since $\sqrt{2} \in \mathbb{F}_{p^2}$, we also seek a square root of $\sigma^2 - \sigma^{-2}$.

To this end, let u denote a point of order 8 on E with $2u = s$. In addition to x, there is an even function y on E satisfying

$$y^2 = 1 - 2\delta x^2 + x^4 .$$

Moreover, y is uniquely fixed by the requirement that $y(0) = 1$. As for x, y is invariant under translation by 2r, but changes sign under translation by 2s. We now apply the duplication formula for the formal group of the Jacobi quartic (cf. (1.2)) to obtain

$$x(2u) = 2x(u)\, y(u)/(1 - x(u)^4) .$$

Write $\zeta = x(u)$ and $\eta = y(u)$, so that

(6.9) $$\sigma = 2\zeta\eta/(1 - \zeta^4) ,$$

since $\sigma = x(s) = x(2u)$.

Resuming the algebra, we use

$$y^2 = 1 - 2\delta x^2 + x^4$$
$$= (1 - \sigma^2 x^2)(1 - \sigma^{-2} x^2)$$

to write

$$\eta^2 = (1 - \sigma^2\zeta^2)(1 - \sigma^{-2}\zeta^2) .$$

Squaring (6.9), it is a simple matter to solve for σ^2 in terms of ζ^2 , namely one has

$$\sigma^2 = 2\zeta^2/(1 + \zeta^4) .$$

From this one obtains the expression

$$\sigma^2 - \sigma^{-2} = (-1)\,\frac{(1-\zeta^4)^2\,\sigma^2}{4\zeta^4} .$$

Since $\sqrt{-1} \in \mathbb{F}_{p^2}$, we shall need to know that <u>both</u> σ <u>and</u> ζ^2 <u>lie in</u> $\mathbb{F}_{p^2}$ to complete the argument.

We now make full use of the assumption that E is supersingular. As in §3, since we have already found that $u_2 = \left(\frac{-1}{p}\right)$ it follows that the p^2-Frobenius map Fr is given by

$$\mathrm{Fr} = \left[\left(\frac{-1}{p}\right) p\right] : E \to E .$$

(Recall that we take E to be defined over $\mathbb{F}_{p^2}$, as noted above.) Hence Fr acts as the identity on $E[4]$, the points of order 4, and acts as either the identity or $[5]$ on $E[8]$.

Now $\mathrm{div}(x)$ is fixed by Fr , hence x^{Fr} and x have the same divisor, and so $x^{\mathrm{Fr}} = cx$ with c a constant. From (6.7) we obtain $c^2 = 1$, since Fr fixes r , and so $x^{\mathrm{Fr}} = \pm x$.

Next observe that from

$$(x(u),y(u)) = (\zeta,\eta)$$

follows

$$(x(5u),y(5u)) = (-\zeta,-\eta) ,$$

since $5u = u + 2s$ and both functions x,y change sign under translation by $2s$. Since Fr sends u to u or to $5u$, and since $x^{\mathrm{Fr}} = \pm x$, we see that in all cases Fr fixes ζ^2,η^2 and $\zeta\eta$. By (6.9), Fr also fixes σ , and so the argument is complete. □

Note that since Fr fixes both $\sigma = x(s)$ and s , with $\sigma \neq 0$, we must have $x^{\mathrm{Fr}} = x$, i.e. x is defined over $\mathbb{F}_{p^2}$.

<u>Appendix</u>: <u>The formal group of a Jacobi quartic</u>. We shall offer an elementary verification of the formula (1.2) due to Euler for the formal group of a Jacobi quartic. Adopting the outlook of §5, it will suffice to verify that the elliptic function

$s(u)$ introduced there satisfies the addition formula

$$s(u+v) = \frac{s(u)\, s'(v) + s'(u)\, s(v)}{1 - \varepsilon s(u)^2\, s(v)^2} \quad . \tag{A.1}$$

We do this by holding v fixed with $s(v) \in \mathbb{C}\setminus\{0\}$, and using divisors to verify that the left and right sides of (A.1) are the same elliptic function of u. This treatment is inspired by the proof of the addition formula for Fueter's elliptic functions given by Cassou-Noguès and Taylor in [23, Ch. IV, §2].

Recall from §5 that $s(u)$ is an odd elliptic function of order 2 with period lattice 2Ω generated by $2\omega_1$ and $2\omega_2$, having divisor

$$\operatorname{div} s(u) = (0) + (\omega_3) - (\omega_1) - (\omega_2) , \tag{A.2}$$

and satisfying $s'(u) = 1$. Moreover, $s(u + \omega_3) = -s(u)$ and so

$$s(\omega_3 - u) = s(u) .$$

Following [23], choose ψ with $2\psi = \omega_3$, e.g. $\psi = \omega_3/2$. Then $s'(u)$ is an even elliptic function of order 4 having the same period lattice, and an easy calculation shows that it has divisor

$$\operatorname{div} s'(u) = (\psi) + (-\psi) + (\psi + \omega_1) + (\psi + \omega_2) - 2(\omega_1) - 2(\omega_2) . \tag{A.3}$$

Next consider $s(u + \omega_1)$. Its divisor is the negative of the divisor of $s(u)$, so

$$s(u + \omega_1)\, s(u) = c \tag{A.4}$$

for a constant $c \neq 0$. Then also

$$s(u + \omega_2)\, s(u) = -c . \tag{A.5}$$

We shall see that $\varepsilon = c^{-2}$. Indeed, we have

$$s'(u)^2 = 1 - 2\delta s(u)^2 + \varepsilon s(u)^4 , \tag{A.6}$$

and so $1 - 2\delta x^2 + \varepsilon x^4 = 0$ has as roots $s(\psi)$, $s(-\psi)$, $s(\psi + \omega_1)$ and $s(\psi + \omega_2)$. Hence

$$1 - 2\delta x^2 + \varepsilon x^4 = \left(1 - \frac{x^2}{s(\psi)^2}\right)\left(1 - \frac{x^2}{s(\psi+\omega_1)^2}\right)$$

and so $\varepsilon = s(\psi)^{-2}\, s(\psi + \omega_1)^{-2} = c^{-2}$, as claimed.

Turning to the addition formula for $s(u)$, we see that it will suffice to verify that

$$s(u+v)\left(1 - \frac{s(u)\, s(v)}{c}\right)\left(1 + \frac{s(u)\, s(v)}{c}\right) = s(u)\, s'(v) + s'(u)\, s(v) . \tag{A.7}$$

For this, we shall fix v with $2v \neq 0$, in $\mathbb{C}/2\Omega$, so that $s(v) \in \mathbb{C}\setminus\{0\}$, and regard (A.7) as an equality between elliptic functions of u. Since (A.7) holds for $u = 0$, it suffices to show that both sides have the identical divisor.

First, from (A.2) we have

(A.8) $$\operatorname{div} s(u + v) = (-v) + (2\psi - v) - (\omega_1 - v) - (\omega_2 - v) .$$

Next, it is an easy exercise to verify that

(A.9) $$\operatorname{div} \left(1 - \frac{s(u)\, s(v)}{c}\right) = (\omega_1 + v) + (\omega_2 - v) - (\omega_1) - (\omega_2)$$

and

(A.10) $$\operatorname{div} \left(1 + \frac{s(u)\, s(v)}{c}\right) = (\omega_2 + v) + (\omega_1 - v) - (\omega_1) - (\omega_2) .$$

Hence the left side of (A.7) is an elliptic function of order 4 with divisor

(A.11) $$(-v) + (2\psi - v) + (\omega_1 + v) + (\omega_2 + v) - 2(\omega_1) - 2(\omega_2) .$$

Finally, we seek the divisor of $s(u)\, s'(v) + s'(u)\, s(v)$ as a function of u. Evidently, since $s(v) \in \mathbb{C}\setminus\{0\}$, it has only the double poles indicated by (A.11), so we need to locate its four zeros. Put $u = -v$ to obtain

$$s(-v)\, s'(v) = s'(-v)\, s(v) = -s(v)\, s'(v) + s'(v)\, s(v) = 0 .$$

Put $u = 2\psi - v$ to obtain

$$s(2\psi - v)\, s'(v) + s'(2\psi - v)\, s(v) = s(v)\, s'(v) - s'(v)\, s(v) = 0 .$$

Put $u = \omega_1 + v$ to obtain

$$s(\omega_1 + v)\, s'(v) + s'(\omega_1 + v)\, s(v)$$

which vanishes: take the derivative of (A.4). In the same way, one takes the derivative of (A.5) to see that

$$s(\omega_2 + v)\, s'(v) + s'(\omega_2 + v)\, s(v) = 0 .$$

We conclude that (A.11) is also the divisor of the right side of (A.7), and so we have verified the addition formula. □

References

1. H. Bateman (A. Erdelyi, ed.): Higher Transcendental Functions, vol. 2, McGraw-Hill, 1953.

2. D.V. Chudnovsky and G.V. Chudnovsky: Elliptic modular functions and elliptic genera, Topology, to appear.

3. D.V. Chudnovsky and G.V. Chudnovsky: letter dated February 6, 1986.

4. D.V. Chudnovsky and G.V. Chudnovsky: Elliptic formal groups over $\mathbb{Z}$ and $\mathbb{F}_p$ in applications to number theory, computer science and topology, in this volume.

5. D.V. Chudnovsky, G.V. Chudnovksy, P.S. Landweber, S. Ochanine and R.E. Stong: Integrality and divisibility of elliptic genera, to appear.

6. E. Copson: An Introduction to the Theory of Functions of a Complex Variable, Oxford Univ. Press, 1935.

7. L. Euler: De integrationis aequationis differentialis $m\,dx/\sqrt{1-x^4} = n\,dy/\sqrt{1-y^4}$, Opera omnia XX (1), 58-79, Teubner-Füssli, 1911-1976.

8. B.H. Gross: letter dated April 7, 1986.

9. J. Igusa: On the transformation theory of elliptic functions, Amer. J. Math. 81 (1959), 436-452.

10. J. Igusa: On the algebraic theory of elliptic modular functions, J. Math. Soc. Japan 20 (1968), 96-106.

11. D. Jackson: Fourier Series and Orthogonal Polynomials, Math. Assoc. Amer., 1941.

12. P.S. Landweber: Elliptic cohomology and modular forms, in this volume.

13. P.S. Landweber, D.C. Ravenel and R.E. Stong: Periodic cohomology theories defined by elliptic curves, to appear.

14. P.S. Landweber and R.E. Stong: Circle actions on Spin manifolds and characteristic numbers, Topology, to appear.

15. S. Lang and H. Trotter: Frobenius Distributions in GL_2-Extensions, Lecture Notes in Math. 504, Springer-Verlag, 1976.

16. A.I. Markushevich: The Remarkable Sine Functions, Elsevier, 1966.

17. S. Ochanine: Sur les genres multiplicatifs définis par des intégrales elliptiques, Topology 26 (1987), 143 - 151.

18. D.C. Ravenel: Complex Cobordism and Stable Homotopy Groups of Spheres, Academic Press, 1986.

19. J.-P. Serre: Propriétés galoisiennes des points d'ordre fini des courbes elliptiques, Invent. Math. 15 (1972), 259-331.

20. J.H. Silverman: The Arithmetic of Elliptic Curves, Graduate Texts in Math. 106, Springer-Verlag, 1986.

21. J.H. Wahab: New cases of irreducibilities for Legendre polynomials, Duke J. Math. 19 (1952), 165-176.

22. E.T. Whittaker and G.N. Watson: A Course of Modern Analysis, fourth edition, Cambridge Univ. Press, 1927.

23. Ph. Cassou-Noguès and M.J. Taylor: Elliptic Functions and Rings of Integers, Birkhauser Boston, 1987.

SOME WEIL GROUP REPRESENTATIONS MOTIVATED BY ALGEBRAIC TOPOLOGY

Jack Morava
Department of Mathematics
The Johns Hopkins University
Baltimore, Maryland 21218

1. Introduction

Elliptic cohomology [6] appears to be related to the one - loop nonlinear sigma - model, much as K - theory turned out to be related to elliptic operator theory . This suggests that other localisations of classical bordism functors may have interesting analytical interpretations, some of these being related to multiloop string physics.

This paper is a report on these other localisations, or at least on some of their aspects; because of the nature of the problem it is written very locally, i.e. p - adically. Decomposing things into their p-local components is not a very familiar procedure to physicists, just as renormalisation is not a technique well-understood by algebraic topologists; but (as Mike Hopkins pointed out to me) they can both be interpreted (in 'simple' cases, such as the Riemann zeta - function) as dual approaches to the problem of analytic continuation.

Very briefly [cf. Ravenel's book [10]], the minimal complex - orientable multiplicative cohomology functors, say from finite complexes to finite-dimensional vector spaces over fields, play the role of points, or prime ideals in some sense, for the stable homotopy category, much as the points of the prime ideal spectrum of a commutative ring can be identified with the homomorphisms from the ring to algebraically closed fields. There is one such 'chromatic prime' for each p^n, where n is a natural number and p is one of the primes known to Euclid; reduction modulo a chromatic prime corresponds to the K - functor of height n at p, with $n \to \infty$ leading to the 'degenerate' nonperiodic K - functors $H^*(-,F_p)$ at the finite primes as well as the unique $H^*(-,Q)$ at the infinite prime. We have the beginnings of a theory of stable homotopy at these chromatic primes, in the work of Devinatz, Hopkins, and Smith [3], but for many questions suggested by elliptic cohomology we would like to understand better how these primes fit together, and in particular how they adjoin the infinite prime: for elliptic cohomology takes values in categories of modules over very global rings, e.g. rings of modular forms, and it becomes natural to ask in particular about the way the $K(n)$'s lift to cohomology functors taking values in categories of modules over discrete valuation rings (e.g. $W(F_p)$) which have quotient fields of characteristic zero. [Over global number rings there is a nice way of understanding an orientable cohomology functor as an Euler product of local factors associated to its p - completions, as in Honda [2,§6].] I should add here that the Proceedings of the 1987 Durham symposium on homotopy theory [LMS Lecture Notes 117, ed. E. Rees and J. D. S. Jones] are a good reference for this material: in particular see the talks of Hopkins, Ravenel, and Ray.

In the next section we sketch the construction of a lifting of $K(n)$ to the integers of a local number field L of degree n over Q_p; in the third section we show that the rational cohomology functor defined by this lifting is naturally a representation of a certain Weil group, roughly the automorphisms of a maximal abelian extension of L, over Q_p. These Weil groups, or more properly their global analogues, play a central role in Langlands's program for the classification of automorphic representations. Their appearance here suggests that the connections between cobordism and physics suggested by elliptic cohomology will extend in some natural way to involve representation theory. The final section is devoted to examples, mostly related to classifying spaces for finite groups.

By now the list of acknowledgements I ought to make for help in this work is almost uncountable: my colleagues Mike Boardman, George Kempf, Jean-Pierre Meyer, Jeff Smith and Steve Wilson at Hopkins; J. Frank Adams and Alan Robinson in England; Mike Hopkins, Haynes Miller, Doug Ravenel, Bob Stong, Friedhelm Waldhausen, Nobuaki Yagita... A long time ago S. P. Novikov whetted my interest in the integral lifting problem for the K(n)'s; at one point in my talk here Noriko Yui called out "too many q's", and I thus learned of a serious error which I hope I have now corrected. But I owe Peter Landweber special thanks, for help over many years: the unwobbling pivot at the center of this conference.

2. Definitions

Let A be the integers of a finite extension L of the p - adic field Q_p; let π be a generator of its maximal ideal, and let $A/(\pi) = k$ be its residue field. The Lubin - Tate group law's classifying homomorphism

$$\pi_* \underline{\underline{MU}} \to A[t]$$

sends CP(i) to $(q/\pi)^j t^i$ or 0, according to whether $i = q^j-1$ or not, with q signifying the cardinality p^f of k. There are other characterisations of this homomorphism, e.g. in terms of the logarithm

$$\log_\pi(T) = \sum_{i \geq 0} \pi^{-i} T^{q^i}$$

or in terms of the formal Mellin transform, or zeta - factor,

$$(1 - \pi^{-1} q^{1-s})^{-1}$$

of the associated group - law; or in terms of its values $v_m(\pi)t^{p^m-1}$ on the Hazewinkel generators of $\pi_*\underline{\underline{MU}}$, given by the recursive formula

$$p^{-1} \sum_{\substack{i\equiv 0 \bmod f \\ m-1\geq i\geq 1}} \pi^{-i/f} v_{m-i}(\pi)^{p^i} = \begin{cases} \pi^{-m/f} & \text{if } m \equiv 0 \bmod f \\ 0 & \text{if not.} \end{cases}$$

Let $I(\pi)$ be the ideal of $\pi_*\underline{\underline{BP}} \otimes A[t]$ generated by the elements

$$V_i(\pi) = v_i - v_i(\pi)t^{p^i-1};$$

algebraic geometers can take t to stand for the Tate motif. The semisimplicial Koszul construction [7] $K_{A[t,t^{-1}]\otimes \underline{\underline{BP}}}(V_i(\pi) \mid i > 0)$ defines a Sullivan - Baas quotient $\underline{\underline{K}}(\pi) = A[t,t^{-1}]\otimes\underline{\underline{BP}}/\underline{\underline{I}}(\pi)$ which lifts $\underline{\underline{K}}(n)$, in that the cofiber of π-multiplication is a copy of $\underline{\underline{K}}(n)\otimes k$; but it is technically a simplification to work with the ($\underline{\underline{BP}}$ - module) spectrum

$$\underline{\underline{BP}}/\underline{\underline{I}}^0(\pi) = K_{A[t,t^{-1}]\otimes\underline{\underline{BP}}}(V_i(\pi) \mid n>i>0)$$

and to define

$$K(\pi)_*(-) = A[t,t^{-1}]_{A\otimes BP} BP/I^0(\pi)_*(-),$$

which turns out to be represented by $\underline{\underline{K}}(\pi)$. We write $\underline{\underline{K}}_L$ for this spectrum and suppress the π-dependence; there is a multiplicative isomorphism

$$W(\bar{k})\otimes_{W(k)} K(\pi_0)^* \cong W(\bar{k})\otimes_{W(k)} K(\pi_1)^*$$

corresponding to choices π_0, π_1 of uniformising element. The main tool below will be the analogues

$$\eta_R(v_i) \equiv pt_i - \sum_{i>j>0} t_j \eta_R(v_{i-j})^{p^j} \quad \bmod (v_1, \ldots v_{n-1})$$

of some congruences of Ravenel.

The case when L is the unramified extension $W(F_q) \otimes Q$, and so f = n, is worth discussing as an example. Taking π to be p, the Todd genus for this theory sends complex projective i-space to $p^{(n-1)j}$ when $i = q^j - 1$, and to zero otherwise, and the lifting is in fact defined over $W(F_p)$; in particular, when n is one we obtain the 2(p-1) - periodic component of p-adically completed complex K - cohomology [1].

By calculating the cooperation rings we will show below that the rationalisation $K_L^*(-) \otimes Q$ is a representation of a Weil group; in fact we will see that the full ring of cohomology endomorphisms of $\underline{\underline{K}}_L$ is an extension of a group algebra of the Weil group, by an exterior algebra of p - torsion elements. To see how these Weil groups arise in this context requires a few more recollections. Assume L to be normal over Q_p; then there is a short exact sequence

$$1 \to \mathrm{Gal}(L^{ab}/L) \to \mathrm{Gal}(L^{ab}/Q_p) \to \mathrm{Gal}(L/Q_p) \to 1$$

of Galois groups defined by the maximal abelian extension L^{ab} of L. Now the group $\mathrm{Gal}(L^{ab}/L)$ is isomorphic to the profinite completion of the multiplicative group $L^\times = Z \times A^\times$ of L, and the extension above is the completion of a generator of the cyclic cohomology group $H^2(\mathrm{Gal}(L/Q_p), L^\times)$ of order n. The dense open subgroup of $\mathrm{Gal}(L^{ab}/Q_p)$ corresponding to $L^\times$ in

$Gal(L^{ab}/L)$ is called the Weil group (of L^{ab} over Q_p, or more colloquially 'of L'), cf. [13].

As above, the case n = 1 is the motivating example. The Weil group 'of Q_p' is the multiplicative group $Z \times \hat{Z}_p^{\times}$ of Q_p: the initial factor completes to the Galois group $\hat{Z} = Gal(\bar{F}_p/F_p)$ of the maximal unramified subfield of L^{ab} [when L is Q_p, L^{ab} is obtained from Q_p by adjoining all roots of 1, and the completion of the maximal unramified subfield is $W(\bar{F}_p) \otimes Q$], while the second factor acts as the Galois group of the (totally ramified) field of pth power roots of unity over Q_p. It is this group which acts naturally on the p - adic completion of K - theory, as the stable p - adic Adams operations; but for some purposes it is convenient to think of the factor Z as acting as well, on $K^*(-) \otimes W(\bar{F}_p)$, as the subgroup of $Gal(\bar{F}_p/F_p)$ generated by the Frobenius endomorphism $\sigma(x) = x^p$. In general we will think of the maximal compact (inertial) subgroup $N^0(L)$ of the Weil group (the notation will be justified below) as acting on $K_L^*(-) \otimes_{W(k)} W(\bar{k})$ equivariantly under the $Gal(\bar{F}_p/F_p)$ - action.

To see the connection with the K(n)'s, we need a summary of their Adams operations. Let D denote the division algebra obtained by adjoining to $L_0 = W(F_q) \otimes Q$, $q = p^n$, an nth root F of p, satisfying the skew commutation rule

$$Fa = a^{\sigma} F, \ a \in L_0.$$

It can be shown that every extension L of degree n over Q_p appears in D as a maximal commutative subfield, and when L/Q_p is normal its Galois group can be identified with the Weyl group of $L^{\times}$ regarded as a maximal torus in $D^{\times}$. The <u>Weil</u> group of L, on the other hand, is the normaliser N(L) of $L^{\times}$, by a theorem of Weil and Shafarevich [14, appendix III]. The maximal compact subgroup S(D) of $D^{\times}$ acts on $K(n)^*(-, \bar{F}_p)$, equivariantly under the Z - action generated by σ

acting as above on $\bar{F}_p$ and as F - conjugation on S(D), cf. [9,§2.1.3].

Reduction modulo p makes it possible to think of endomorphisms of $\underline{\underline{K}}_L$ as specialising to endomorphisms of K(n)*(-,k), and conversely to interpret endomorphisms of the latter functor as lifting to endomorphisms of $\underline{\underline{K}}_L$.

The technicalities are being reserved for the next section, the generalities for this one; so it is convenient to note here that a key part of the argument below is consideration of the bilateral Hopf algebra

$$A \;\substack{\longrightarrow\\ \longrightarrow\\ \longleftarrow}\; \underline{\underline{C}} = A\otimes_{BP}BP_*BP\otimes_{BP}A \;\substack{\longrightarrow\\ \longrightarrow\\ \longrightarrow\\ \longleftarrow\\ \longleftarrow}\; \cdots$$

which represents the functor assigning to an A - algebra B, the groupoid of automorphisms of the Lubin - Tate - Honda group law defined over B. When B is restricted to be a flat A - algebra, this is the standard isotropy - group of the Lubin - Tate law, and when B is restricted to be a k - algebra we get the isotropy group of its mod π reduction. We will see below that if $C(N^0(L),A)$ denotes the Hopf algebra of continuous A - valued functions on $N^0(L)$, and $C^\infty(S(D),k)$ denotes the similar algebra of k - valued locally constant functions on S(D), then $\underline{\underline{C}} \otimes_A k$ is isomorphic to $C^\infty(S(D),k)$, while $\underline{\underline{C}}$ modulo its ideal of torsion elements is isomorphic to $C(N^0(L),A)$: both rings represent the same functor, in either case. [For simplicity we ignore the $Gal(\bar{F}_p/F_p)$ - action.]

The principal result of the next section can be now displayed as the commutative diagram

$$\begin{array}{ccc} K_{L*}\underline{\underline{K}}_L & \xrightarrow{\sim} & E'_*(\underline{Q}^0)\otimes\underline{\underline{C}} \\ \downarrow & & \downarrow \\ k\otimes K(n)_*\underline{\underline{K}}(n)\otimes k & \xrightarrow{\sim} & E_*(\underline{Q})\otimes C^{\infty}(S(D),k) \end{array}$$

in which the vertical arrows are defined by reduction mod π. The undefined terms on the right are exterior algebras or their quotients which will be constructed in the course of the proof, associated to representations $\underline{Q}^0$(resp.$\underline{Q}$) of $N^0(L)$ (resp. $S(D)$) defined over $W(k)$ (resp. k), which are closely related to the subalgebra of Bockstein operations in the Steenrod algebra. The argument will show that the algebra $E'_*(\underline{Q}^0)$ is torsion in positive dimensions; this then verifies the assertion made above, that $K_L^*(-)\otimes Q$ applied to a finite complex yields a finite - dimensional representation of a Weil group. More precisely, there is a $Gal(\bar{F}_p/F_p)$ - equivariant action of $N^0(L)$ on the $W(\bar{k})\otimes Q$ - vector space defined by the rationalisation of $K_L^*(-)\otimes_{W(k)}W(\bar{k})$.

3. The Exterior Representations

Applying $BP/I^0(\pi)$ to the filtered $\underline{\underline{BP}}$ - module spectrum $\underline{\underline{BP}}/\underline{I}^0(\pi)$, following Johnson and Wilson [5], we form the differential graded Koszul algebra

$$K^*_{BP/I^0(\pi)_*\underline{\underline{BP}}}(\eta_R(V_i(\pi))\mid n>i>0),$$

where

$$\eta_R(V_i(\pi)) = \eta_R(v_i) - v_i(\pi)t^{p^i-1} \in BP/I^0(\pi)_*\underline{\underline{BP}};$$

its homology is the E_2 - term of a spectral sequence converging to an algebra of cooperations on $BP/I^0(\pi)$, but this algebra is not in general flat over $BP/I^0(\pi)_*$, since $\underline{\underline{BP}}/\underline{I}^0(\pi)\wedge\underline{\underline{BP}}/\underline{I}^0(\pi)$ is not usually a wedge of suspensions of $\underline{\underline{BP}}/\underline{I}^0(\pi)$.

Now $v_i(\pi) \equiv 0 \bmod \pi$ $(n>i>0)$ in general, but when $L = L_0$ is unramified we can take π to be p and $v_i(\pi)$ to be zero$(n>i>0)$; the graded algebra above then simplifies to (the pullback over $W(F_q)$ of the) $W(F_p)$-algebra

$$K^*_{v_n^{-1}BP/I_n^0{}_*\underline{\underline{BP}}}(\eta_R(v_i) \mid n>i>0),$$

$I_n^0 = (v_i | n>i>0)$. By the Ravenel - style congruences cited above, the elements $x_i = p^{-1}\eta_R(v_i)$, $n>i>0$, are p - integral and in fact define polynomial generators $W(F_p)[t_1,\ldots,t_{n-1}]$; moreover,

$$x_i \equiv t_i \mod (t_1,\ldots,t_{i-1}),\ n>i>0,$$

so the sequence $(\eta_R(v_1),\ldots,\eta_R(v_{n-1}))$ will be regular only if n is two or less, and in general this differential algebra is not a resolution of its homology

$$(v_n^{-1}BP/I_n^0)[\underline{\underline{t}}]/(pt_i | n>i>0)$$

in degree zero.

However in this case it is not hard to work out the homology directly. We can write

$$K^*_{v_n^{-1}BP/I_n^0[\underline{\underline{t}}]}(px_i | n>i>0) =$$

$$E_*(\underline{\underline{Q}}_i^0 | n>i>0)\otimes(v_n^{-1}BP/I_n^0)[x_i | n>i>0]\ [t_n,t_{n+1},\ldots];$$

the differential [cf. Serre, [11, IV §2]] is multiplication by the class

$$p\underline{Q}^0x = \sum_{n>i>0} \underline{Q}_i^0\eta_R(v_i).$$

[In the general case the differential is multiplication by $p\underline{Q}^0\tilde{x} + \pi a$, where $a = \sum a_i\underline{Q}_i^0$ is an A - linear sum of exterior generators and $\tilde{x}_i$ is analogous to x_i; the A - module structure of the E_2 - term then looks pretty forbidding.]

Now the sequence x_i $(n>i>0)$ is regular, so if the graded algebra above had the differential $\underline{Q}^0x$ it would be acyclic, and we could identify the kernel of $\underline{Q}^0x$ - multiplication with its cokernel. But ker $p\underline{Q}^0x$ = ker $\underline{Q}^0x$, while im $p\underline{Q}^0x$ = p im $\underline{Q}^0x$, and the homology of the complex with differential $p\underline{Q}^0x$ is the reduction mod p, in positive dimensions, of the cokernel of $\underline{Q}^0x$.

If V is a free graded $W(F_p)$ - module, let $E_*(V)$ denote its exterior algebra and write $\tilde{E}_*(V) = E_*(V)/pE^+(V)$ for its reduction mod p in positive dimensions. The homology of our Koszul algebra can be displayed as the product

$$((\tilde{E}_*(\underline{Q}_i^0|n>i>0) \otimes ((v_n^{-1}BP/I_n^0)[x_i|n>i>0]/(px_i)))/(\underline{Q}^0x))[t_i|i\geq n],$$

which can be simplified further since $x_i \equiv t_i$ mod p.

Tensoring this complex over BP with A on both sides yields the product of an exterior algebra with the algebra $\underline{C}$ of §2; moreover this tensored complex is rationally acyclic, since $\eta_R(V_i(\pi))$ is regular in $BP/I^0(\pi)_*\underline{\underline{BP}}\otimes Q$. Reduction modulo π maps $BP/I^0(\pi)_*\underline{\underline{BP}}/\underline{I}^0(\pi)$ to

$$BP/(p,I_n^0)_*\underline{\underline{BP}}/(p,I_n^0) \cong E_*(\underline{Q})\otimes(BP/(p,I_n^0))[\underline{t}]$$

sending $\underline{\underline{Q}}_i^0$ to $\underline{\underline{Q}}_i$; to identify the associated isotropy representations (i.e. after tensoring over BP with A) is our next task. The mod p representations were constructed in [9,§2.1.5] from the mod p reduction homomorphism

$$W(F_q) < F > / (F^n - p) \longrightarrow F_q < F > / (F^n).$$

The group of units of the ring on the left is S(D); the homomorphism thus defines an action of S(D) on the F_q - vector space spanned by the elements $\underline{\underline{Q}}_i = F^{i-1}$, n>i>0, equivariant under $Gal(F_q/F_p)$. On the other hand $N^0(L)$ is a subgroup of the units of D, and hence acts (by right multiplication) on the maximal ideal of its valuation ring, spanned by elements $\underline{\underline{Q}}_i^0 = F^{i-1}$, n>i>1, to define a representation over $W(F_q)$, Galois - equivariant as above.

We conclude that $K_L^*(-)\otimes Q$ is a comodule over the rationalisation of the tensor product of an exterior algebra of torsion elements, with $\underline{\underline{C}}$; but $\underline{\underline{C}} \otimes Q$ is just a (somewhat twisted) group algebra of $N^0(L)$.

4. Examples

This is all fairly complicated even in the simplest cases. For example, $\tilde{K}_L^0(S^2)$ is a free A - module on one generator; the abelian subgroup $A^\times$ of $N^0(L)$ acts by multiplication. In general, a in $A^\times$ acts on $\tilde{K}_L^0(S^{2m})$ by a^m-multiplication. Thus K_2^L is isomorphic, as $Gal(L^{ab}/L)$ - module, to the Tate module of the Lubin - Tate grouplaw, and $K_L^*(pt) \cong A[t,t^{-1}]$ as graded modules, with $[a](t) = at$. More generally, $N^0(L)$ acts on $\tilde{K}_L^0(S^{2m})\otimes_{W(k)} W(\bar{k})$, defining a representation which seems to be hard to construct directly, cf. [8,§4].
As a final example consider $K_L(BZ/pZ) \cong A[[T]]/(T^{-1}[p](T))$.

This is the Hopf algebra representing the groupscheme of p - torsion points on the Lubin - Tate group; it is the union of its subgroupschemes of π - torsion points, and

$$K_L(BZ/pZ)\otimes Q = L \oplus L_0 \oplus \ldots \oplus L_{e-1}, \quad e = n/f,$$

is isomorphic to the sum of L with the fields L_i generated over it by the π^{i+1} - torsion points. It is classical that $[L_i:L] = (q-1)q^i$.

We conclude with some (even more) speculative remarks, suggested by the extraordinary character theory developed recently by Hopkins, Kuhn, and Ravenel. If G is a finite group, we can define the set of its one parameter A - module subgroups

$$\hat{G}(A) = \mathrm{Hom}_c(A,G)/G$$

to be the set of continuous homomorphisms, up to conjugacy, from A to G. Let

$$\hat{R}_{\bar{L}}(G) = C(\hat{G}(A),\bar{L})$$

be the functions on $\hat{G}(A)$ with values in an algebraic closure $\bar{L}$ of L, made a $\mathrm{Gal}(\bar{L}/L)$ - module by the action

$$f^\sigma(\phi) = [\sigma^{-1}]f(\sigma\phi), \quad \sigma \in \mathrm{Gal}(\bar{L}/L), \ \phi \in \hat{G}(A)$$

[generalising the action in Serre [12,§12.4]], where $\hat{G}(A)$ is a Galois - module courtesy of the Artin homomorphism

$$\mathrm{Gal}(\bar{L}/L) \longrightarrow \mathrm{Gal}(L^{ab}/L) \xrightarrow{\sim} \hat{Z}\times A^\times \longrightarrow A^\times .$$

It seems reasonable to conjecture, motivated by Quillen's approach to algebraic K - theory, that $K^0_L(-)\otimes_A\bar{L}$ is the representable envelope of the functor

$$X \longrightarrow \hat{R}_{\bar{L}}(\pi_1(X))$$

in the category of profinite spaces.

REFERENCES

[1] M. F. Atiyah, D. O. Tall, Group representations, λ - rings and the J - homomorphism, Topology 8 (1969), 253 - 297.

[2] T. Honda, Theory of commutative formal groups, J. Math. Soc. Japan 22(1970), 213 - 245.

[3] M. Hopkins, E. Devinatz, J. Smith, Nilpotence in stable homotopy theory I, preprint, to appear in Annals of Math.

[4] M. Hopkins, N. Kuhn, D. Ravenel, work in progress.

[5] D. Johnson, W. S. Wilson, BP - operations and extraordinary K - theories, Math Zeits. 144(1975), 55 - 75.

[6] P. Landweber, Elliptic cohomology and modular forms, in these proceedings.

[7] J. Morava, A product for the odd - primary bordism of manifolds with singularities, Topology 18(1979), 177 - 187.

[8] -----, The Weil group as automorphisms of the Lubin - Tate group, Astérisque 63(1979), 169 - 178.

[9] -----, Noetherian localisations of categories of cobordism comodules, Annals of Math. 121(1985), 1 - 39.

[10] D. Ravenel, Complex Cobordism and Stable Homotopy Groups of Spheres, Academic Press, 1986.

[11] J. P. Serre, Algèbre Locale: multiplicités, Lecture Notes in Math 11, Springer-Verlag, 1965.

[12] -----, Linear Representations of Finite Groups, Graduate Texts in Math. 42, Springer-Verlag, 1977.

[13] A. Weil, Sur la théorie des corps de classes, J. Math. Soc. Japan 3(1951) 1 - 35.

[14] -----, Basic Number Theory, Grundl. Math Wiss. Bd. 144, Springer-Verlag (3rd edition, 1974).

GENRES ELLIPTIQUES EQUIVARIANTS

Serge Ochanine
CNRS, Université de Paris-Sud, Mathématique
91405 Orsay, France

Les genres elliptiques sont nés d'une tentative de comprendre, en termes topologiques, une conjecture de E.Witten [13] concernant l'indice équivariant de l'opérateur de Rarita-Schwinger sur une variété spinorielle munie d'une action d'un groupe de Lie compact G. Il se trouve que cet indice peut être inclus dans un genre elliptique universel construit par P.Landweber et R.Stong [10]. La conjecture initiale devient ainsi la conséquence d'une conjecture plus générale concernant les genres elliptiques. Sous cette forme élaborée, la question a été traitée récemment par E.Witten [14] dans le cadre de la théorie quantique des champs. A défaut d'une preuve rigoureuse de la conjecture, [14] fournit néanmoins un fort argument heuristique en sa faveur.

Le présent article a pour but limité d'introduire la notion de genre elliptique équivariant nécessaire à l'énoncé de la conjecture et de démontrer celle-ci dans un cas relativement simple - celui où l'action du groupe G est semi-libre. Le résultat principal dit, essentiellement, que dans le cas où G est le cercle, le genre elliptique équivariant est une fonction elliptique dont les pôles ne peuvent occuper que des positions bien précises sur le tore de définition, ne dépendant que des nombres de rotation de la variété. Des résultats plus forts, mais toujours incomplets et techniquement beaucoup plus lourds, peuvent être obtenus par une méthode dérivée de [7,8], dans le cas, notamment, où la variété admet une structure faiblement complexe invariante par G. L'exposé de ces résultats dépasserait le cadre de cet article.

L'intérêt constant et les nombreux commentaires et suggestions de Peter Landweber et Bob Stong m'ont beaucoup aidé. Je suis heureux de leur exprimer ici toute ma reconnaissance.

§1. Genres équivariants.

Dans ce paragraphe, Λ désigne une $\mathbb{C}$-algèbre commutative unitaire.

1.1. Cohomologie équivariante (cf.[1],§2). Soit G un groupe de Lie compact connexe et $EG \longrightarrow BG$ un G-fibré universel. Pour tout G-espace M, on pose:

$$M_G = EG \times_G M.$$

C'est un espace fibré de base BG et de fibre M. On notera p sa projection.

La cohomologie équivariante de M est définie par:

$$H_G^*(M) = H^*(M_G;\Lambda).$$

Par exemple,

$$H_G^*(pt) = H^*(BG;\Lambda),$$

et, pour $G = 1$,

$$H_G^*(M) = H^*(M;\Lambda).$$

La projection p induit un homomorphisme

$$p^*: H_G^*(pt) \longrightarrow H_G^*(M)$$

qui fait de $H_G^*(M)$ un $H_G^*(pt)$-module.

Lorsque M est une variété compacte orientée, p induit également un homomorphisme de Gysin

$$p_*: H_G^*(M) \longrightarrow H_G^*(pt)$$

(ou "intégration sur la fibre", cf.[3],§8) qui est un morphisme de $H_G^*(pt)$-modules gradués, homogène de degré $-\dim M$.

Soit $f: K \longrightarrow G$ un morphisme de groupes de Lie. Alors tout G-espace M est également un K-espace via f, et on a un homomorphisme

$$f^{\#}: H_G^*(M) \longrightarrow H_K^*(M)$$

qui commute avec p^* et, lorsque M est une variété orientée, avec p_*.

Par exemple, le morphisme $1 \longrightarrow G$ induit un homomorphisme d'augmentation

$$a: H_G^*(M) \longrightarrow H^*(M;\Lambda)$$

qui, pour $M = pt$, donne un isomorphisme

$$a: H_G^0(pt) \xrightarrow{\simeq} \Lambda.$$

On notera $\overline{H}_G^*(pt)$ l'idéal d'augmentation, i.e. le noyau de a pour

$M = pt$.

Pour tout espace Z, on pose:

$$\hat{H}^*(Z;\Lambda) = \prod_{i\geq 0} H^i(Z;\Lambda),$$

et pour tout G-espace M:

$$\hat{H}^*_G(M) = \hat{H}^*(M_G;\Lambda).$$

On peut démontrer (cf. [12], théorème 2.1) que si M est un complexe fini, $H^*_G(M)$ est un $H^*_G(pt)$-module de type fini. Il s'ensuit que dans ce cas $\hat{H}^*_G(M)$ est la complétion $\bar{H}^*_G(pt)$-adique de $H^*_G(M)$, ce qui explique la notation $\hat{H}^*_G(M)$.

Comme les morphismes p^*, p_* et $f^{\#}$ sont homogènes, ils s'étendent naturellement aux complétions.

Si V est un G-fibré vectoriel orienté de dimension m sur M, on définit

$$V_G = EG \times_G V.$$

C'est un fibré vectoriel orienté sur M_G. Pour toute classe caractéristique $\chi \in \hat{H}^*(BSO_m;\Lambda)$, on pose:

$$\chi_G(V) = \chi(V_G) \in \hat{H}^*_G(M).$$

Ce sont les classes caractéristiques équivariantes de V, liées aux classes habituelles par:

$$\chi(V) = a\,\chi_G(V).$$

Plus généralement, pour tout morphisme $f: K \longrightarrow G$, le G-fibré V est aussi un K-fibré, et on a:

$$f^{\#}\chi_G(V) = \chi_K(V).$$

Dans la suite, on attachera une importance particulière au cas où $G = S$, le cercle unité de $\mathbb{C}$. On peut choisir l'espace projectif $\mathbb{C}P^\infty$ pour BS, et l'on a:

$$H^*_S(pt) = \Lambda[u],$$

où $u \in H^2(\mathbb{C}P^\infty)$ est le générateur canonique. Si M est un complexe fini, alors $\hat{H}^*_S(M)$ est la complétion u-adique de $H^*_S(M)$. En particulier:

$$\hat{H}^*_S(pt) = \Lambda[[u]],$$

l'anneau des séries formelles sur Λ, et l'augmentation

$$a: \hat{H}^*_S(pt) \longrightarrow \Lambda$$

associe à une série formelle son terme constant.

1.2. Genres équivariants. Soit Ω_*^{SO} l'anneau de cobordisme orienté, et

$$\varphi : \Omega_*^{SO} \longrightarrow \Lambda$$

un Λ-genre, i.e. un morphisme d'anneaux unitaires. On notera

$$g(u) = \sum_{i \geq 0} \frac{\varphi\,[\mathbb{C}P^{2i}]}{2i+1} u^{2i+1}$$

le logarithme de φ, et $\Phi \in \hat{H}^*(BSO; \Lambda)$ la classe totale de Hirzebruch associée, qui est une classe multiplicative entièrement définie par sa valeur sur le fibré en droites universel L sur $\mathbb{C}P^\infty$:

$$\Phi(L) = u/g^{-1}(u).$$

On a alors:

$$\varphi[M] = \Phi(M)[M] = \Phi(TM)[M],$$

où TM est le fibré tangent de M.

Par exemple, le Â-genre est défini par son logarithme

$$g(u) = 2 \text{ arésinh } (u/2)$$

ou par

$$\hat{A}(L) = \frac{u}{2 \sinh(u/2)} .$$

Soit M une G-variété compacte orientée de dimension k. Le fibré TM est canoniquement un G-fibré.

Définition: $\varphi_G[M] = p_* \Phi_G(TM) \in \hat{H}_G^*(pt)$.

Comme $\Phi_G(TM)$ est concentré dans les dimensions divisibles par 4, $\varphi_G[M]$ est concentré dans les dimensions congrues à $-k$ modulo 4. Par exemple, pour $G = S$, la série

$$\varphi_S[M] \in \Lambda[[u]]$$

est nulle pour $k \equiv 1 \bmod 2$; paire pour $k \equiv 0 \bmod 4$, et impaire pour $k \equiv 2 \bmod 4$.

Si $f: K \longrightarrow G$ est un morphisme de groupes de Lie, on a:

$$f^\# p_* \Phi_G(TM) = p_* f^\# \Phi_G(TM) = p_* \Phi_K(TM),$$

donc

$$f^\# \varphi_G[M] = \varphi_K[M].$$

En particulier, pour $K = 1$, on obtient:

$$\varphi_G[M] \equiv \varphi[M] \quad \bmod \bar{H}_G^*(pt).$$

On dira que $\varphi_G[M]$ est <u>constant</u>, si

$$\varphi_G[M] = \varphi[M].$$

<u>Proposition 1</u>. Soit φ un Λ-genre. Les conditions suivantes sont équivalentes:

(i) pour toute S-variété spinorielle M, $\varphi_S[M]$ est constant;

(ii) pour tout groupe de Lie compact connexe G et toute G-variété spinorielle M, $\varphi_G[M]$ est constant;

(iii) pour tout G-fibré principal $E \longrightarrow B$, où B est une variété compacte orientée, et toute G-variété spinorielle M, on a:

$$\varphi[E \times_G M] = \varphi[B]\, \varphi[M].$$

<u>Démonstration</u>. (i) $\Longrightarrow$ (ii). Soit G un groupe de Lie et M une G-variété spinorielle. Si $i: T \longrightarrow G$ est l'inclusion d'un tore maximal, le morphisme

$$i^{\#} : \hat{H}^*_G(pt) \longrightarrow \hat{H}^*_T(pt)$$

est injectif. Donc $\varphi_T[M] = i^{\#}\, \varphi_G[M]$ est constant si et seulement si $\varphi_G[M]$ est constant. On se réduit ainsi au cas où $G = T$.

On a:

$$H^*_T(pt) = \Lambda[u_1,\dots,u_m], \quad m = \dim T,$$
$$H^*_S(pt) = \Lambda[u].$$

Un morphisme $f: S \longrightarrow T$ est déterminé par un vecteur $(n_1,\dots,n_m) \in \mathbb{Z}^m$:

$$f^{\#} u_i = n_i u.$$

Si $h(u_1,\dots,u_m)$ est un élément non-nul de $H^k_T(pt)$, on a:

$$f^{\#}\, h(u_1,\dots,u_m) = h(n_1,\dots,n_m)u^k \neq 0$$

pour un choix approprié de f. Donc, si $\varphi_T[M]$ est non-constant, il existe un $f: S \longrightarrow T$ tel que $f^{\#}\, \varphi_T[M]$ soit non-constant.

(ii) $\Longrightarrow$ (iii). Prenons un G-fibré principal $E \longrightarrow B$ et soit

$$b: B \longrightarrow BG$$

l'application classifiante. Il lui correspond un diagramme de fibrés:

$$\begin{array}{ccc} E \times_G M & \xrightarrow{\ \bar{b}\ } & M_G \\ {\scriptstyle q}\downarrow & & \downarrow{\scriptstyle p} \\ B & \xrightarrow{\ b\ } & BG \end{array}$$

On remarque que $\bar{b}^*(TM)_G = E \times_G TM$ est le fibré des vecteurs tangents aux fibres de q. Donc

$$T(E \times_G M) = q^*TB \oplus (E \times_G TM).$$

Si $\varphi_G[M] = \varphi[M]$, on a:

$$\begin{aligned}\varphi[E \times_G M] &= q_*\Phi(E \times_G M)[B] \\ &= (\Phi(B)q_*\Phi(E \times_G TM))[B] \\ &= (\Phi(B)b^* \varphi_G[M])[B] \\ &= \varphi[B]\,\varphi[M].\end{aligned}$$

(iii) $\Rightarrow$ (i). Soit M une S-variété spinorielle et

$$\varphi_S[M] = \varphi[M] + \alpha u^k + o(u^k) \qquad (\alpha \in \Lambda).$$

Considérons le S-fibré principal $S^{2k+1} \longrightarrow \mathbb{C}P^k$. L'application classifiante coïncide avec l'inclusion canonique

$$b: \mathbb{C}P^k \longrightarrow \mathbb{C}P^\infty.$$

Comme précédemment, on obtient:

$$\begin{aligned}\varphi[S^{2k+1} \times_S M] &= (\Phi(\mathbb{C}P^k)b^* \varphi_S[M])[\mathbb{C}P^k] \\ &= \varphi[\mathbb{C}P^k]\,\varphi[M] + \alpha.\end{aligned}$$

Donc (iii) entraîne $\alpha = 0$, d'où le résultat. □

<u>Remarques</u>. 1. En termes de [3],§21, la condition (ii) signifie que la classe Φ est <u>strictement multiplicative</u> dans les fibrés $M_G \longrightarrow BG$, où M est une G-variété spinorielle.

2. La proposition est applicable au $\hat{A}$-genre, puisque d'après [2], $\hat{A}_S[M] = 0$ (cf. 2.3 ci-dessous).

3. La proposition reste vraie si l'on remplace partout "variété spinorielle" par "variété orientée". La propriété multiplicative de la signature τ (cf.[5]) implique alors que $\tau_G[M]$ est constant pour toute G-variété orientée M.

4. Il y a deux cas extrêmes où il est facile de voir que $\varphi_G[M]$ est constant. Si l'action de G sur M est <u>triviale</u>, $\varphi_G[M]$ est de la forme

$$f^\# \varphi_1[M] = \varphi[M],$$

où f est le morphisme $G \longrightarrow 1$. A l'opposé, si G est un tore et si l'action de G sur M est <u>libre</u>, on sait (cf.[1], corollaire 3.6) que $H_G^*(M)$ est un $H_G^*(pt)$-module de torsion. Il en résulte que $\varphi_G[M] = 0$.

1.3. <u>Genres elliptiques</u>. Rappelons qu'un Λ-genre φ est dit elliptique, si son logarithme est donné par une intégrale formelle:

$$g(u) = \int_0^u \frac{dz}{\sqrt{R(z)}}, \tag{1}$$

où $R(z) = 1 - 2\delta z^2 + \varepsilon z^4 \quad (\delta, \varepsilon \in \Lambda)$.

On démontre (cf. [11], théorème 1) que φ est elliptique si et seulement si pour tout fibré vectoriel complexe V de dimension paire sur une variété compacte orientée, on a $\varphi[\mathbb{C}P(V)] = 0$, où $\mathbb{C}P(V)$ désigne la projectivisation de V.

Proposition 2 (R.Stong). Tout Λ-genre φ satisfaisant aux conditions équivalentes de la proposition 1 est elliptique.

Démonstration. Soit $E \longrightarrow B$ un U_{2k}-fibré principal et

$$V = E \times_{U_{2k}} \mathbb{C}^{2k}$$

le fibré vectoriel associé. Le groupe U_{2k} opère canoniquement sur la variété spinorielle $\mathbb{C}P^{2k-1}$ et l'on a:

$$\mathbb{C}P(V) = E \times_{U_{2k}} \mathbb{C}P^{2k-1} = E/U_{2k-1}.$$

Si φ vérifie (iii), on a:

$$\varphi[\mathbb{C}P(V)] = \varphi[B]\,\varphi[\mathbb{C}P^{2k-1}] = 0$$

pour des raisons de dimension. Donc φ est elliptique. □

Conjecture. Les conditions suivantes sont équivalentes:

(a) le genre φ est elliptique;

(b) $\varphi_G[M]$ est constant pour tout groupe de Lie G et toute G-variété spinorielle M.

En raison des propositions 1 et 2, cette conjecture est équivalente à l'affirmation: pour toute S-variété spinorielle M et tout genre elliptique φ, le genre équivariant $\varphi_S[M]$ est constant. De plus, la remarque 4 ci-dessus montre qu'il suffit de considérer le cas où l'ensemble M^S des points fixes de S est non-vide. C'est sous cette forme que la conjecture sera abordée dans le paragraphe suivant.

§2. Genre elliptique universel.

2.1. Fonction elliptique x. Soit W un réseau arbitraire de $\mathbb{C}$ et

$$r: W \longrightarrow \{\pm 1\}$$

un homomorphisme surjectif.

Proposition 3. Il existe une fonction méromorphe unique x sur $\mathbf{C}$ telle que

(i) x est impaire;

(ii) les pôles de x sont précisément les points de W; ils sont tous simples et le résidu de x en $w \in W$ est $r(w)$;

(iii) pour tout $w \in W$, on a:

$$x(u + w) = r(w)x(u).$$

Démonstration. Unicité. Si x_1 et x_2 sont deux fonctions méromorphes satisfaisant à (i)-(iii), et $f = x_1 - x_2$, on a:

(i) f est impaire;

(ii) f n'a pas de pôle dans $\mathbf{C}$;

(iii) f est une fonction elliptique relative au réseau $W_0 = \ker(r)$.

Il en résulte que f est une fonction constante impaire, donc nulle.

Existence (cf. [15]). Soit w_1, w_2 une base de W telle que

$$r(w_1) = 1, \quad r(w_2) = -1.$$

Posons

$$x(u) = \sum_{w \in W} \frac{r(w)}{u + w} \tag{2}$$

$$= \sum_{m \in \mathbf{Z}} \Big(\sum_{n \in \mathbf{Z}} \frac{(-1)^n}{u + mw_1 + nw_2} \Big)$$

$$= \sum_{m \in \mathbf{Z}} \frac{\pi}{w_2 \cdot \sin\left(\frac{\pi(u+mw_1)}{w_2}\right)} .$$

La dernière de ces séries est absolument et localement uniformément convergeante dans $\mathbf{C} \setminus W$, et définit bien une fonction méromorphe satisfaisant à (i)-(iii). $\square$

Remarque. Comme la fonction x est entièrement déterminée par ses propriétés (i)-(iii), la fonction construite ne dépend pas du choix de la base w_1, w_2 de W. Ceci justifie l'expression (2) pour x.

Il est facile de voir, comme cela se fait pour la fonction $\wp$ de Weierstraß et sa dérivée, que si

$$x(u) = u^{-1} + g_2^* u + g_4^* u^3 + \ldots \tag{3}$$

est le développement de Laurent de x en $u = 0$, la fonction x et sa dérivée y sont liées par

$$y^2 = x^4 - 2\delta x^2 + \varepsilon \tag{4}$$

où

$$\delta = -3g_2^*, \qquad \varepsilon = 7g_2^{*2} - 10g_4^*.$$

Soit φ le $\mathbb{C}$-genre elliptique défini par l'intégrale (1) avec pour δ et ε les valeurs indiquées. Alors la série $g^{-1}(u)$ est le développement en $u = 0$ de $1/x$ et la classe de Hirzebruch correspondante est déterminée par

$$\Phi(L) = 1 + g_2^* u^2 + g_4^* u^4 + \dots$$

(cf. 1.2).

2.2. Genre elliptique universel (cf.[9,10,4,15]). Soit

$$H = \{z \in \mathbb{C} \mid \operatorname{Im}(z) > 0\}$$

le demi-plan de Poincaré. Pour $z \in H$, considérons le réseau W_z engendré par $4\pi iz$ et $2\pi i$. Les formules

$$r(4\pi iz) = 1, \qquad r(2\pi i) = -1$$

définissent une surjection $r\colon W_z \longrightarrow \{\pm 1\}$. On notera $u \longmapsto x(u,z)$ la fonction x correspondante.

Soit $\Gamma = SL_2(\mathbb{Z})$ le groupe modulaire et $\Gamma_0(2)$ le sous-groupe de Γ des matrices

$$\begin{pmatrix} a & b \\ c & d \end{pmatrix}$$

telles que $c \equiv 0 \bmod 2$. Le groupe Γ, et donc aussi $\Gamma_0(2)$, opèrent sur H par

$$\begin{pmatrix} a & b \\ c & d \end{pmatrix} z = \frac{az + b}{cz + d}$$

et l'on vérifie facilement (cf.[15]) que pour un élément de $\Gamma_0(2)$, on a:

$$x\left(\frac{u}{cz + d}, \frac{az + b}{cz + d}\right) = (cz + d)\,x(u,z),$$

ou encore,

$$g_{2k}^*\left(\frac{az + b}{cz + d}\right) = (cz + d)^{2k} g_{2k}^*(z)$$

pour les coefficients du développement de Laurent (3).

On peut démontrer que $g_{2k}^*(z)$ $(k \geq 1)$ est une forme modulaire de poids $2k$ pour $\Gamma_0(2)$ et que si $M_*(\Gamma_0(2))$ est l'anneau gradué de telles formes, on a:

$$M_*(\Gamma_0(2)) = \mathbf{C}[g_2^*, g_4^*] = \mathbf{C}[\delta, \varepsilon].$$

En conséquence, le genre φ de 2.1 peut être considéré comme genre multiplicatif sur $M_*(\Gamma_0(2))$, et il a la propriété universelle évidente suivante: pour tout Λ-genre elliptique φ_1, il existe un morphisme unique

$$\lambda : M_*(\Gamma_0(2)) \longrightarrow \Lambda$$

tel que $\varphi_1 = \lambda\varphi$.

Il est clair qu'il suffit de démontrer la conjecture pour ce genre universel φ, ou, ce qui est équivalent, pour tout genre elliptique φ construit dans 2.1 à partir d'un réseau W et d'une surjection r.

2.3. Conjecture de Witten. Soit M une variété riemannienne compacte de dimension 2m munie d'une structure spinorielle, et soit

$$\mathcal{S} = \mathcal{S}^+ \oplus \mathcal{S}^-$$

le fibré des spineurs sur M. Pour tout fibré vectoriel réel V sur M muni d'une métrique riemannienne et d'une connexion orthogonale, il existe un opérateur de Dirac

$$D_V^+ : \Gamma(\mathcal{S}^+ \otimes_{\mathbf{R}} V) \longrightarrow \Gamma(\mathcal{S}^- \otimes_{\mathbf{R}} V)$$

dont l'indice

$$\operatorname{ind} D_V^+ = \dim \ker D_V^+ - \dim \operatorname{coker} D_V^+$$

est donné par la formule

$$\operatorname{ind} D_V^+ = ph(V)\, \hat{A}(M)[M]$$

(cf.[6],§1), où $ph(V) = ch(\mathbf{C} \otimes_{\mathbf{R}} V)$.

Par exemple, l'opérateur de Dirac "classique" D^+ correspond au cas où V est le fibré trivial de dimension 1, et on a:

$$\operatorname{ind} D^+ = \hat{A}[M].$$

Un autre exemple est donné par l'opérateur de Rarita-Schwinger qui correspond à V = TM. On a:

$$\operatorname{ind} D_{TM}^+ = ph(TM)\, \hat{A}(M)[M].$$

Supposons maintenant qu'un groupe de Lie compact connexe G opère par isométries, de manière compatible, sur M, sur le $Spin_{2m}$-fibré principal qui définit la structure spinorielle sur M, et sur V (dans ce cas l'action de G est de type pair - cf.[2],§2). Alors D_V^+ est un G-opérateur et il a un indice équivariant

$$\operatorname{ind}_G D_V^+ = [\ker D_V^+] - [\operatorname{coker} D_V^+] \in R(G),$$

où $R(G)$ est l'anneau des représentations complexes de G. Soit

$$\mathrm{ch}\colon R(G) \longrightarrow \hat{H}^*(BG)$$

le morphisme qui à chaque G-module complexe U associe le caractère de Chern du fibré associé sur BG:

$$U \longmapsto \mathrm{ch}(EG \times_G U).$$

Alors le théorème de l'indice pour les familles d'opérateurs elliptiques permet d'affirmer que

$$\mathrm{ch}(\mathrm{ind}_G D_V^+) = p_*(\mathrm{ph}_G(V)\, \hat{A}_G(M)).$$

Par exemple, on a:

$$\hat{A}_G[M] = \mathrm{ch}(\mathrm{ind}_G D^+) = 0$$

d'après [2].

Dans [13], E.Witten énonce la conjecture suivante:

Conjecture de Witten. L'indice équivariant de l'opérateur de Rarita-Schwinger est constant.

Nous allons voir comment cette conjecture découle de la conjecture du §1.

Soit

$$\varphi\colon \Omega_*^{SO} \longrightarrow M_*(\Gamma_0(2))$$

le genre elliptique universel de 2.2, et

$$\lambda\colon M_*(\Gamma_0(2)) \longrightarrow \mathbf{C}[[q]]$$

le morphisme qui à chaque forme modulaire associe sa série de Fourier en $q = \exp(2\pi i z)$. On notera $\varphi^q = \lambda\varphi$ le $\mathbf{C}[[q]]$-genre elliptique correspondant et Φ^q la classe de Hirzebruch de φ^q.

Proposition 4. On a:

$$\varphi_G^q[M] = -\mathrm{ch}(\mathrm{ind}_G D_{TM}^+)q + o(q).$$

Démonstration. Les coefficients de Fourier de $x(u,z)$ sont explicitement calculés dans [15]. On a, en particulier,

$$ux(u,z) = \frac{u}{2\sinh(u/2)}\,(1 - (e^u + e^{-u} - 2)q + o(q)).$$

Donc, pour tout fibré vectoriel orienté V de dimension $2m$, on a:

$$\Phi^q(V) = \hat{A}(V)(1 - \mathrm{ph}(V - 2m)q + o(q))$$

et

$$\varphi_G^q[M] = (1 + 2mq)\,\hat{A}_G[M] - p_*(\mathrm{ph}_G(TM)\cdot \hat{A}_G(M)) + o(q)$$

$$= -\mathrm{ch}(\mathrm{ind}_G D^+_{TM})q + o(q)$$

(dim M = 2m). □

Remarque. Les autres coefficients de Fourier de $\varphi^q_G[M]$ s'interprètent également comme des indices équivariants d'opérateurs de Dirac appropriés. Voir à ce sujet la formule de Witten pour le genre elliptique universel ([14],cf.[9],§6).

2.4. Formule d'intégration (cf.[1],§3). Soit M une S-variété compacte orientée et F une composante connexe de l'ensemble M^S des points fixes, supposé non-vide. Soit N^F le fibré normal de F dans M. Le cercle S opère dans les fibres de N^F sans vecteur non-nul fixe. Il en découle qu'il existe une décomposition unique de N^F en facteurs isotypiques

$$N^F = \bigoplus_{j \in J(F)} N^F_j,$$

où $J(F)$ est un ensemble fini d'entiers positifs, et où N^F_j est un fibré vectoriel complexe sur lequel $z \in S$ opère par multiplication par z^j. On oriente F de manière compatible avec les orientations de M et de N^F. On pose $J(M) = \bigcup_F J(F)$. Ce sont les nombres de rotation de M.

Comme l'action de S sur F est triviale, on a:

$$F_S = BS \times F$$

et

$$N^F_S = \bigoplus_{j \in J(F)} (L^j \otimes N^F_j),$$

où L^j est la puissance tensorielle j-ième du fibré en droites universel sur BS. Si E désigne la classe d'Euler, il est clair que la composante de

$$E_S(N^F) = \prod_{j \in J(F)} E(L^j \otimes N^F_j)$$

dans $H^*(BS) \otimes H^0(F)$ est égale à

$$\prod_{j \in J(F)} (ju)^{\dim_{\mathbb{C}} N^F_j}.$$

Comme les éléments de degré positif de $H^*(F)$ sont nilpotents, $E_S(N^F)^{-1}$ est un élément bien défini de $H^*_S(F)[u^{-1}]$.

Proposition 5 (cf. [1],§3). Pour tout $h \in H^*_S(M)$, on a dans $H^*_S(pt)[u^{-1}]$:

$$p_* h = \sum_F p^F_*(E_S(N^F)^{-1} i^*_F h),$$

où F parcourt les composantes connexes de M^S, i_F est l'inclusion canonique de F dans M et

$$p_*^F\colon H_S^*(F) \longrightarrow H_S^*(pt)$$

est le morphisme de Gysin relatif à F. □

2.5. Application au calcul de $\varphi_S[M]$. Soit, comme dans 2.1, W un réseau de $\mathbf{C}$ et $r\colon W \longrightarrow \{\pm 1\}$ une surjection. On notera $\mathcal{E}$ le corps des fonctions elliptiques relatives au réseau $W_0 = \ker(r)$, et $\mathcal{O}$ l'anneau local en $u = 0$, i.e. le sous-anneau de $\mathcal{E}$ des fonctions régulières en ce point.

Pour tout $n \in \mathbf{Z}$, soit $\mathcal{E}_n$ le sous-espace des fonctions $f \in \mathcal{E}$ telles que

$$f(u + w) = r(w)^n f(u)$$

pour tout $w \in W$. Il est clair que $\mathcal{E}_n$ ne dépend que de n modulo 2 et qu'on a $\mathcal{E} = \mathcal{E}_0 \oplus \mathcal{E}_1$.

De même, on pose $\mathcal{O}_n = \mathcal{O} \cap \mathcal{E}_n$ et l'on a $\mathcal{O} = \mathcal{O}_0 \oplus \mathcal{O}_1$. Il est clair que les fonctions de $\mathcal{O}_n$ sont régulières en tout point de W.

L'application qui à chaque fonction f associe sa série de Laurent en $u = 0$, définit des plongements

$$\mathcal{E} \longrightarrow \hat{H}_S^*(pt)[u^{-1}]$$
$$\mathcal{O} \longrightarrow \hat{H}_S^*(pt),$$

et l'on identifiera toujours $\mathcal{E}$ et $\mathcal{O}$ avec leurs images.

Les fonctions x et y appartiennent à $\mathcal{E}_1$. Plus généralement, soit $T \subset SO_{2m}$ le tore maximal standard, et considérons dans

$$\hat{H}^*(BT)[u_1^{-1},\ldots,u_m^{-1}] = \mathbf{C}[[u_1,\ldots,u_m]][u_1^{-1},\ldots,u_m^{-1}]$$

le produit

$$X = x(u_1)\ \ldots\ x(u_m).$$

Il est clair que X est invariant par l'action du groupe de Weyl de SO_{2m} et représente donc une classe

$$X \in \hat{H}^*(BSO_{2m})[E^{-1}]$$

telle que

$$X \equiv E^{-1} \mod \hat{H}^*(BSO_{2m}).$$

Proposition 6. Soit φ le genre elliptique de 2.1. Alors on a:

$$\varphi_S[M] = \sum_F p_*^F(X_S(N^F)\ \Phi_S(F)).$$

Démonstration. La classe Φ est définie par

$$\Phi(L) = ux(u).$$

Donc

$$\Phi_S(N^F) = E_S(N^F)X_S(N^F),$$

d'où

$$i_F^* \Phi_S(M) = \Phi_S(N^F)\,\Phi_S(F) = E_S(N^F)X_S(N^F)\,\Phi_S(F),$$

et le résultat découle de la proposition 5. □

Lemme. Pour tout $k \geq 1$, la dérivée k-ième $x^{(k)}$ de x est un polynôme à coefficients complexes en x et y.

Démonstration. L'affirmation est évidente pour $k = 1$, puisque $x' = y$.

Pour $k = 2$, on a $x'' = y'$. En dérivant (4), on obtient:

$$2yy' = 4x^3y - 4\delta xy,$$

d'où

$$y' = 2x^3 - 2\delta x.$$

Enfin, si

$$x^{(k-1)} = Q(x,y),$$

où Q est un polynôme à coefficients complexes, on a:

$$x^{(k)} = Q_1(x,y)y + Q_2(x,y)(2x^3 - 2\delta x),$$

où Q_1 et Q_2 sont les dérivées partielles de Q. Le lemme en découle par récurrence. □

Soit M une S-variété spinorielle connexe telle que $M^S \neq \emptyset$. Alors on sait (cf.[2], 2.4) que les nombres

$$n(F) = \sum_{j \in J(F)} j \dim_{\mathbf{C}} N_j^F$$

ne dépendent pas, modulo 2, de la composante F: ils sont tous pairs, si la variété est de type pair, et tous impairs, si la variété est de type impair.

Proposition 7. Soit M une S-variété compacte orientée. Alors

(i) $\varphi_S[M] \in \mathcal{O}$ est un polynôme en $x(ju)$ et $y(ju)$, où j parcourt $J(M)$;

(ii) si M est spinorielle et connexe, on a:

$$\varphi_S[M] \in \mathcal{O}_n$$

où n est la parité de l'action.

Démonstration. On a:

$$X_S(N^F) = \prod_{j \in J(F)} X(L^j \otimes N_j^F).$$

Soient v_{ij} $(j \in J(F);\ i = 1,\dots,\dim_{\mathbb{C}} N_j^F)$ des variables telles que la classe de Chern totale de N_j^F soit donnée par

$$c(N_j^F) = \prod_i (1 + v_{ij}).$$

Alors

$$X_S(N^F) = \prod_j \prod_i x(ju + v_{ij}).$$

La formule de Taylor

$$x(u + v) = x(u) + x'(u)v + \dots$$

et le lemme montrent alors que

$$X_S(N^F) = \sum f_k \otimes h_k,$$

où $h_k \in H^*(F)$ et où $f_k \in \mathcal{E}_{n(F)}$ est un polynôme en $x(ju)$ et $y(ju)$ $(j \in J(F))$. La proposition en découle. □

Cette proposition montre que les pôles de $\varphi_S[M]$ sont tous <u>d'ordre fini modulo W</u>, plus précisément, que tout pôle est annulé, modulo W, par un nombre de rotation de la variété.

<u>Théorème</u>. Soit Λ une $\mathbb{C}$-algèbre commutative unitaire et φ un Λ-genre elliptique. Alors pour toute S-variété spinorielle semi-libre M, le genre équivariant $\varphi_S[M]$ est constant.

<u>Démonstration</u>. D'après 2.2, il suffit de démontrer le théorème pour le genre φ construit dans 2.1 à partir d'un réseau W et d'une surjection r . La proposition 7 est alors applicable.

Comme l'action de S est semi-libre, S opère librement dans les fibres de N^F. Donc tous les nombres de rotation sont égaux à 1 et $\varphi_S[M]$ est un polynôme en x et y. Il s'ensuit que les seuls pôles possibles de $\varphi_S[M]$ sont les points de W, ce qui est exclu puisque $\varphi_S[M] \in \mathcal{O}_n$.

<u>Références</u>.

1. M.F.Atiyah et R.Bott: The moment map and equivariant cohomology. Topology 23(1984), 1-28.

2. M.F.Atiyah et F.Hirzebruch: Spin-manifolds and group actions. In: Essays on Topology and Related Topics, Springer-Verlag,1970, 18-28.

3. A.Borel et F.Hirzebruch: Characteristic classes and homogeneous spaces, I, II. Amer. J. Math. 80(1958), 458-538; 81(1959), 315-382.

4. D.V.Chudnovsky et G.V.Chudnovsky: Elliptic modular functions and elliptic genera. Topology, à paraître.

5. S.S.Chern, F.Hirzebruch et J.-P.Serre: On the index of a fibered manifold. Proc. Amer. Math. Soc. 8(1957), 587-596.

6. M.Gromov et H.B.Lawson,Jr.: Positive scalar curvature and the Dirac operator on complete Riemannian manifolds. Publ. Math. I.H.E.S. 58(1983), 83-196.

7. I.M.Kričever: Formal groups and Atiyah-Hirzebruch formulae. Izv. Akad. Nauk SSSR, Ser.Mat. 38(1974), 1289-1304.

8. I.M.Kričever: Obstructions for the existence of S^1-actions. Bordism of branched coverings. Izv. Akad. Nauk SSSR, Ser.Mat. 40(1976), 828-844.

9. P.S.Landweber: Elliptic cohomology and modular forms. Dans ce volume.

10. P.S.Landweber et R.E.Stong: Circle actions on Spin manifolds and characteristic numbers. Topology, à paraître.

11. S.Ochanine: Sur les genres multiplicatifs définis par des intégrales elliptiques. Topology, à paraître.

12. D.Quillen: The spectrum of an equivariant cohomology ring, I, II. Annals of Math. 94(1971), 549-602.

13. E.Witten: Fermion quantum numbers in Kaluza-Klein theory. In: Proceedings of the 1983 Shelter Island conference on quantum field theory and the fundamental problems of physics (ed. R.Jackiw et al.) MIT Press 1985, 227-277.

14. E.Witten: Elliptic genera and quantum field theory. Communications in Mathematical Physics, à paraître.

15. D.Zagier: Note on the Landweber-Stong elliptic genus. Dans ce volume.

Complex Cobordism Theory for Number Theorists

DOUGLAS C. RAVENEL

Department of Mathematics
University of Washington
Seattle, WA 98195

§1. Elliptic cohomology theory

The purpose of this paper is to give the algebraic topological background for elliptic cohomology theory and to pose some number theoretic problems suggested by these concepts.

Algebraic topologists study functors from the category of spaces to various algebraic categories. In particular there are functors to the category of graded rings called multiplicative generalized cohomology theories. (All rings are assumed to be commutative and unital. Graded rings are commutative subject to the usual sign conventions, i.e., odd-dimensional elements anticommute with each other.) These functors satisfy all of the Eilenberg-Steenrod axioms but the dimension axiom. In other words they have the same formal properties as ordinary cohomology except that the cohomology of a point may be more complicated.

For a discussion of these axioms the interested reader should consult [S] or [ES]; for generalized cohomology theories a good reference is Part III of [A1].

Among these cohomology theories the following examples are mentioned elsewhere in this volume:

Examples 1.1

(i)	$H^*(\cdot; R)$	ordinary cohomology with coefficients in a ring R.
(ii)	$K^*(\cdot)$	complex K-theory.
(iii)	$KO^*(\cdot)$	real K-theory.
(iv)	$MU^*(\cdot)$	complex cobordism theory.
(v)	$MSO^*(\cdot)$	oriented cobordism theory.
(vi)	$MSpin^*(\cdot)$	Spin cobordism theory.

A comprehensive reference for cobordism theory is [St]; a very brief account can be found in Sections 4.1 and 4.2 of [R1].

Partially supported by the National Science Foundation

Typeset by $\mathcal{A}_{\mathcal{M}}\mathcal{S}$-TEX

A cohomology theory E is said to be **complex-oriented** if it behaves well on infinite dimensional complex projective space $\mathbb{C}P^\infty$, namely if

$$E^*(\mathbb{C}P^\infty) = E^*(pt.)[[x]] \text{ with } x \in E^2(\mathbb{C}P^\infty).$$

Here $E^*(pt.)$ means the cohomology of a point and x is such that its restriction to $E^*(pt.)$ is zero. This is a graded ring which will be abbreviated by E^*. For any space X, $E^*(X)$ is an algebra over E^*. Of the examples given above, all but (iii) and (vi) are complex-oriented. Complex-oriented theories are studied in detail in Part II of [A1] and in Sections 4.1 and 4.2 of [R1].

When the theory is complex-oriented, there is a formal group law associated with it.

DEFINITION 1.2. *A* **formal group law** *over a commutative ring with unit R is a power series $F(x, y)$ over R that satisfies*

(i) $F(x,0) = F(0,x) = x$ (identity),
(ii) $F(x,y) = F(y,x)$ (commutativity) and
(iii) $F(F(x,y),z) = F(x,F(y,z))$ (associativity).

(The existence of an inverse is automatic. It is the power series $i(x)$ determined by the equation $F(x,i(x)) = 0$.)

Examples 1.3

(i) $F(x,y) = x + y$. This is called the additive formal group law.
(ii) $F(x,y) = x + y + xy = (1+x)(1+y) - 1$. This is called the multiplicative formal group law.
(iii)

$$F(x,y) = \frac{x\sqrt{R(y)} + y\sqrt{R(x)}}{1 - \varepsilon x^2 y^2}$$

where

$$R(x) = 1 - 2\delta x^2 + \varepsilon x^4.$$

This is the formal group law associated with the elliptic curve

$$y^2 = R(x),$$

a Jacobi quartic. It is defined over $\mathbb{Z}[1/2][\delta, \varepsilon]$. This curve is nonsingular mod p (for p odd) if the discriminant $\Delta = \varepsilon(\delta^2 - \varepsilon)^2$ is invertible. This example figures prominently in elliptic cohomology theory; see [L1] for more details.

The theory of formal group laws from the power series point of view is treated comprehensively in [Ha]. A short account containing all that is relevant for the current discussion can be found in Appendix 2 of [R1].

If E^* is an oriented cohomology theory then we have

$$E^*(CP^\infty \times CP^\infty) = E^*(pt.)[[x \otimes 1, 1 \otimes x]].$$

There is a map from $CP^\infty \times CP^\infty$ to CP^∞ inducing the tensor product of complex line bundles. The induced map in cohomology goes the other way since cohomology is a contravariant functor. Thus we get a map

$$E^*(pt.)[[x]] \to E^*(pt.)[[x \otimes 1, 1 \otimes x]].$$

DEFINITION 1.4. *Let $F_E(x \otimes 1, 1 \otimes x)$ denote the image of x under this map. It is easy to verify that this F_E satisfies the conditions of 1.2, so we have a formal group law over the ring E^*.*

Now we need a result from the theory of formal group laws first proved by Lazard.

THEOREM 1.5. *(i) There is a universal formal group law defined over a ring L of the form*

$$G(x,y) = \sum_{i,j} a_{i,j} x^i y^j \quad \text{with } a_{i,j} \in L$$

such that for any formal group law F over R there is a unique ring homomorphism θ from L to R such that

$$F(x,y) = \sum_{i,j} \theta(a_{i,j}) x^i y^j.$$

(ii) L is a polynomial algebra $\mathbb{Z}[x_1, x_2, \ldots]$. If we put a grading on L such that $a_{i,j}$ has degree $2(1-i-j)$ then x_i has degree $-2i$.

The relevance of this to algebraic topology is embodied in the following result of Quillen [Q] proved in 1969.

THEOREM 1.6. *The formal group law F_{MU} associated with complex cobordism theory is isomorphic to Lazard's universal formal group law. In particular L is isomorphic to MU^*.*

This suggests that complex cobordism should be central to the study of oriented theories and their relation to formal group laws. On the other hand Ochanine's theorem [O] concerns oriented manifolds, which suggests using oriented cobordism theory MSO^*. This discrepancy is not a serious one. Every complex manifold is oriented, so there is a natural homomorphism from MU^* to MSO^*. Moreover if we localize away from the prime 2, i.e., if we tensor with $\mathbb{Z}[1/2]$, then MU^* and MSO^* are essentially equivalent theories, i.e., each one is functorially determined by the other.

Quillen's theorem (1.6) also raises the question of whether there is an oriented cohomology theory for each formal group law. A formal group law over R corresponds by 1.5 to a homomorphism from L to R, i.e., an L-module structure on R. One can ask if there is an MU-module spectrum E which is a ring spectrum such that $E^* = R$ with the desired L-module structure. There are no known counterexamples, but also no general theorems. Given such an R, a more precise question is whether the functor

$$R^*(X) = MU^*(X) \otimes_L R,$$

is a cohomology theory. It is if it satisfies certain criteria spelled out in the Landweber Exact Functor Theorem [L2]; a precise statement can be found in 4.2 of [R1].

The formal group law of 1.3 (ii) does satisfy Landweber's criteria while 1.3 (i) does not. In the case of (ii) the resulting cohomology theory is classical complex K-theory. We get elliptic cohomology from (iii) after inverting the prime 2 and any of $\delta^2 - \varepsilon$, ε or Δ. The proof in the latter case involves some deeper aspects of formal group law theory and elliptic curves; see [LRS]. In particular it uses the fact the the formal group law of an elliptic curve must have height 1 or 2; this is proved as Corollary 7.5 in Silverman's book [Si]. (The height of a formal group law is defined below in 2.4.)

Unfortunately this method of definition is too abstract to reveal the full power of the cohomology theory in question. In the case of K-theory there is the classical definition in terms of vector bundles, which bears little obvious relation with complex cobordism or formal group laws. Any 'physical' interpretation of K-theory, such as its relation to the Dirac operator, relies completely on the geometric definition and would not be possible without it.

It would be extremely desirable to have a comparable geometric definition of elliptic cohomology. In [Wi] Witten studies a possible geometric definition of the map from MSO^* to R used to define elliptic cohomology.

§2. Some curious group cohomology and the chromatic filtration

In studying complex cobordism theory, topologists have been led to the study of the cohomology (in the algebraic sense of the term) of certain groups which appear to be arithmetically interesting.

DEFINITION 2.1. *Let Γ be the group of power series over $\mathbb{Z}$ having the form*

$$\gamma = x + b_1x^2 + b_2x^3 + \cdots$$

where the group operation is functional composition. Γ *acts on the Lazard ring* L *of 1.5 as follows. Let* $G(x,y)$ *be the universal formal group law as above and let* $\gamma \in \Gamma$. *Then* $\gamma^{-1}G(\gamma(x),\gamma(y))$ *is another formal group law over* L, *and therefore is induced by a homomorphism from* L *to itself. Since* γ *is invertible, this homomorphism is an automorphism, giving the desired action of* Γ *on* L.

The cohomology group in question is $H^*(\Gamma; L)$. It is of topological interest because it is closely related to the homotopy groups of spheres. More precisely, it is the E_2-term of the Adams-Novikov spectral sequence, which converges to the stable homotopy groups of spheres. For more discussion of this point, see Sections 1.3 and 1.4 of [R1]. The group is bigraded since L itself is graded. It is known that $H^0(\Gamma; L) = \mathbf{Z}$ (concentrated in dimension 0) and that $H^s(\Gamma; L)$ is locally finite for $s > 0$.

This group is difficult to compute and a lot of machinery has been developed for doing so. $H^{1,*}$ and $H^{2,*}$ are completely known. The problem can be be studied locally at each prime p. For $s > 2$, the p-component of $H^{s,t}$ is known for $t < 2(p-1)p^3$ for p odd and for $t < 40$ for $p = 2$. The description of H^1 is very suggestive.

THEOREM 2.2. *$H^{1,t}$ is trivial if t is odd and is cyclic of order 2 if t is a positive odd multiple of 2. When $t = 4k, H^{1,t}$ is cyclic of order a_{2k} (for $k = 1$ its order is half this number or 12), where a_{2k} is the denominator of $B_{2k}/4k$, and B_{2k} is the $2k^{th}$ Bernoulli number.*

The values of a_{2k} for small k are displayed in the following table.

k	1	2	3	4	5	6	7	8	9	10	11	12
a_{2k}	24	240	504	480	264	65520	24	16320	28728	13200	552	131040

These numbers can be described in at least two different ways. $B_{2k}/4k$ is the value of the Riemann zeta function $\zeta(1-2k)$. As such it appears as a coefficient in an Eisenstein series associated with the Weierstrass $\wp$-function. An alternate description which is more useful for our purposes is that

$$a_{2k} = \gcd n^L(n^{2k} - 1) \tag{2.3}$$

where n ranges over all the integers and L is as large as necessary. (This is proved is [A2].) For $2k = 2$, the square of any odd integer is congruent to 1 mod 8, so whenever n is even we take $L \geq 3$ and this makes a_2 divisible by 8. The square of any integer prime to 3 is congruent to 1 mod 3, so when n is divisible by 3 we take $L \geq 1$ and a_2 is divisible by 3. There are no other such congruences for larger primes so $a_2 = 24$ as indicated. For k odd 2.3 gives $a_k = 2$, which is consistent with 2.2.

The groups $H^{2,t}$ are also known. They are far more complicated than $H^{1,t}$; for example there is no upper bound on the number of summands the p-component can have. Moreover there is no known arithmetic description comparable to 2.2. Given the fact that the group has such a complicated structure, a number theoretic interpretation could lead to a great deal of new information.

There is a deeper way to view $H^*(\Gamma; L)$ due to Jack Morava. $Spec(L)$, which is an infinite-dimensional affine space, can be though of as a moduli space for formal group laws over $\mathbf{Z}$. Then the orbits under Γ are isomorphism classes of formal group laws. The classification of formal group laws becomes quite manageable if we replace $\mathbf{Z}$ by $\overline{\mathbf{F}}_p$, the algebraic closure of the field with p elements. We lose little information about cohomology since there is an isomorphism

$$H^*(\Gamma; L \otimes \mathbf{Z}/p) \otimes \overline{\mathbf{F}}_p \cong H^*(\Gamma_{\overline{\mathbf{F}}_p}; L \otimes \overline{\mathbf{F}}_p)$$

where $\Gamma_{\overline{\mathbf{F}}_p}$ is the group of power series over $\overline{\mathbf{F}}_p$ similar to Γ.

Formal group laws over $\overline{\mathbf{F}}_p$ are determined up to isomorphism by an invariant called the **height**. To define it we introduce some power series associated with a formal group law. For each integer n define $[n](x)$ (called the ***n*-series**) by

$$\begin{aligned} [1](x) &= x, \\ [n](x) &= F(x, [n-1](x)) \qquad \text{for } n > 1 \text{ and} \\ [-n](x) &= i([n](x)). \end{aligned} \tag{2.4}$$

These satisfy

$$\begin{aligned} [n](x) &\equiv nx \mod (x^2), \\ [m+n](x) &= F([m](x), [n](x)) \qquad \text{and} \\ [mn](x) &= [m]([n](x)). \end{aligned}$$

In characteristic p the p-series always has leading term ax^q where $q = p^h$. The height is defined to be h. For the additive formal group law we have $[p](x) = 0$ and the height is said to be ∞. The multiplicative formal group law has height 1 since $[p](x) = x^p$. The mod p reduction (for p odd) of the elliptic formal group law of 1.3(iii) has height one or two depending on the values of δ and ε. For example if $\delta = 0$ and $\varepsilon = 1$ then the height is one for $p \equiv 1 \bmod 4$ and 2 for $p \equiv 3 \bmod 4$. (See pp.373-374 of [R1].)

THEOREM 2.5. *(i) For each prime p there are polynomial generators v_i of of $L \otimes \mathbf{Z}_{(p)}$ having degree $2p^i - 2$ such that the height of a formal group law induced by θ is n if and only if*

$$\theta(v_i) = 0 \qquad \text{for } i < n \text{ and } \theta(v_n) \neq 0.$$

(ii) Two formal group laws over $\overline{\mathbf{F}}_p$ *are isomorphic if and only if they have the same height.*

(iii) Let $I_n = (p, v_1, \ldots v_{n-1}) \subset L$. *Then*

$$H^*(\Gamma; v_n^{-1}L/I_n) \otimes \overline{\mathbf{F}}_p \cong H^*(S_n; \overline{\mathbf{F}}_p)$$

where S_n *is the automorphism group of a height* n *formal group law over* $\overline{\mathbf{F}}_p$.

The structure of S_n is known and is described in Chapter 6 of [R1]. It is a group of units in a division algebra D_n (with Hasse invariant $1/n$) over the p-adic numbers $\mathbf{Q}_p$ of rank n^2. It is known that each degree n extension of $\mathbf{Q}_p$ embeds as a subfield of D_n.

The isomorphism 2.5(iii) is very useful since the cohomology of S_n is much easier to compute than that of Γ. This can be used to get information about $H^*(\Gamma; L)$ with the help of the chromatic spectral sequence, which is described in Chapter 5 of [R1]. The cohomology groups of 2.5(iii) are "v_n-periodic" in the sense that they are modules over the ring

$$K(n)^* = \mathbf{Z}/p[v_n, v_n^{-1}]. \tag{2.6}$$

This notion can be generalized in such a way that each element in $H^*(\Gamma; L)$ is v_n-periodic, and this leads to a chromatic filtration of $H^*(\Gamma; L)$.

We know now that this filtration can be defined on the homotopy category itself, not just on $H^*(\Gamma; L)$. The author made several conjectures concerning this in [R2], many of which have been proved by Devinatz, Hopkins and Smith in [DHS]. Expository accounts can be found in [R3] and [Ho]. We will describe some of this material now.

The ring $K(n)^*$ of 2.6 is the coefficient ring of a cohomology theory called the nth **Morava K-theory.** We use $K(0)^*$ to denote rational cohomology. $K(1)^*$ is mod p complex K-theory. Every graded module over $K(n)^*$ is free, so we can regard $K(n)^*$ as a graded field. In particular there is a very convenient Künneth isomorphism

$$K(n)^*(X \times Y) \cong K(n)^*(X) \otimes_{K(n)^*} K(n)^*(Y).$$

Ordinary cohomology with coefficients in a field (such as $\mathbf{Z}/p$ or $\mathbf{Q}$) enjoys a similar property. In [DHS] it is shown that ordinary cohomology with field coefficients and the various Morava K-theories are the only theories with such a Künneth isomorphism. If we tensor with the field with p^n elements, then we get a functorial action of the group S_n of 2.5(iii). For $n = 1$ this group is the p-adic units and we get the p-adic Adams operations.

If G is a finite group then we can define a $K(n)$-theoretic generalization of group cohomology by considering $K(n)^*(BG)$, where BG is the classifying space of G. (Recall that the usual Eilenberg-Mac Lane cohomology of G is $H^*(BG)$.) Unlike $H^*(BG), K(n)^*(BG)$

is known to have finite rank. Atiyah showed that $K^*(BG)$ is the completion of the complex representation ring of G at its augmentation ideal, so $K(1)^*(BG)$ can be described in similar terms. In particular its rank is the number of conjugacy classes of elements in G whose order is a power of p. In [HKR] we generalize this result as follows. The Euler characteristic (and presumably the rank) of $K(n)^*(BG)$ is the number of conjugacy classes of n-tuples of elements in G which commute with each other and have order a power of p.

Morava K-theory is also useful for studying the general homotopy theory of finite complexes. If X is a finite complex it is known that

$$rk\ K(n)^*(X) \leq rk\ K(n+1)^*(X).$$

These numbers are all finite and not all zero (unless X is contractible). After fixing a prime p we say that a finite complex has **type** n if n is the smallest integer such that $K(n)^*(X) \neq 0$. Equivalently n is the smallest integer such that $MU^*(X)$ is not annihilated by the invariant prime ideal I_n of 2.5(iii).

THEOREM 2.7 (MITCHELL [M]). *For each n there is a finite complex X_n of type n.*

It is known that for any finite complex X of type n there is an Γ-equivariant L-endomorphism α of $MU^*(X)$ which becomes an isomorphism after tensoring with $K^*(n)$. One such α is multiplication by an appropriate power of v_n. This is an algebraic property of L-modules with Γ-action. It raises the geometric question of the existence of an analogous endomorphism of X itself.

PERIODICITY THEOREM 2.8 (HOPKINS-SMITH [HS]). *Any finite complex of type n admits an endomorphism α which induces a $K(n)^*$- isomorphism. Moreover some iterate of α is in the center of the endomorphism ring $End(X)$, which has Krull dimension one.*

It is also known that any other Γ-equivariant endomorphism of $MU^*(X)$ is nilpotent, i.e., some iterate of it is zero. The analogous geometric fact is the following.

NILPOTENCE THEOREM 2.9 (DEVINATZ-HOPKINS-SMITH [DHS]). *An endomorphism of a finite complex that induces the trivial map in each Morava K-theory is nilpotent.*

The special case of 2.9 when X is the sphere spectrum is Nishida's Theorem, which says that each positive-dimensional element in the stable homotopy ring is nilpotent.

§3. Formal A-modules

DEFINITION 3.1. *Let A be the ring of integers in a number field K (or a subring thereof) or in a finite extension of the p-adic numbers. A* **formal A-module** *over an A-algebra R*

is a formal group law over R equipped with power series $[a](x)$ *for each* $a \in A$ *with similar properties to the* $[n](x)$ *of 2.4.*

Lubin and Tate [LT] used a formal A-module in the local case to construct the maximal totally ramified abelian extension of the local field K. This is a generalization of Kronecker's Jugendtraum, which concerns the case when K is an imaginary quadratic extension of $\mathbf{Q}$. There we have an elliptic curve E with complex multiplication whose endomorphism ring is A. The formal group law associated with E (over A) is a formal A-module. In both cases the abelian extensions are obtained by adjoining the roots of $[a](x)$.

The theory of formal A-modules is treated in Section 21 of [Ha].

It is possible to generalize the algebraic constructions of the previous section to formal A-modules. There is a generalization of Lazard's theorem (1.5). The resulting ring is denoted by L_A. There are inclusions

(3.2) $$L \otimes A \subset L_A \subset L \otimes K$$

where K is the field of fractions of A. L_A is known to be a polynomial ring in the local case and when the class number of K is 1.

We define the group Γ_A to be the group of power series over A analogous to Γ (2.1). It acts on L_A in a similar way, so we can ask about $H^*(\Gamma_A; L_A)$. It is known that $H^0(\Gamma_A; L_A) = A$ (concentrated in dimension 0) and that $H^s(\Gamma_A; L_A)$ is locally finite for $s > 0$. The topological significance of this group is unclear. It could be the E_2-term of the Adams-Novikov spectral sequence for some generalization of the sphere, but this appears to be very hard to prove. Theorem 2.5 generalizes to formal A-modules in a satisfactory way as does the chromatic spectral sequence. More details can be found in [R4]. In the local case if q is the cardinality of the residue field then we have a sparseness result,

(3.3) $$H^{s,t}(\Gamma_A; L_A) = 0 \quad \text{unless } t \text{ is divisible by } 2q-2.$$

$H^1(\Gamma_A; L_A)$ has been computed in the local case by Keith Johnson [Jo]. The two descriptions of this group (2.2 and 2.3) is the case $A = \mathbf{Z}$ generalize in different ways. In the global case one might generalize 2.2 by using the Dedekind zeta function for K, but this will not work since this function vanishes at negative integers unless K is totally real, while $H^1(\Gamma_A; L_A)$ is far from trivial.

However we can generalize 2.3 by defining ideals

(3.4) $$J_k = \cap (a^N (a^k - 1))$$

where the intersection is over all $a \neq 0 \in A$ and all natural numbers N. Then Johnson's result is that in the local case

(3.5) $$H^{1,2k}(\Gamma_A; L_A) = A/J_k$$

except for certain small values of k and certain rings A. (Recall that for $A = \mathbb{Z}$ there was an exception for $k = 2$.)

References

[A1] J. F. Adams, *Stable Homotopy and Generalized Homology*, Univ. of Chicago Press, 1974.

[A2] J. F. Adams, On the groups J(X), IV, *Topology* 5(1966), 21-71.

[DHS] E. Devinatz, M. J. Hopkins and J. H. Smith, Nilpotence and stable homotopy theory, to appear.

[ES] S. Eilenberg and N. E. Steenrod, *Foundations of Algebraic Topology*, Princeton Univ. Press, 1952.

[Ha] M. Hazewinkel, *Formal Groups and Applications*, Academic Press, 1978.

[Ho] M. J. Hopkins, Global methods in homotopy theory, to appear in Proceedings of the Durham Symposium on Homotopy Theory 1985, Cambridge University Press, 1987.

[HKR] M. J. Hopkins, N. Kuhn and D. C. Ravenel, Morava K-theory and generalized characters for finite groups, to appear.

[HS] M. J. Hopkins and J. H. Smith, Periodicity and stable homotopy theory, to appear.

[Jo] K. Johnson, The Conner-Floyd map for formal A-modules, to appear.

[L1] P. S. Landweber, Elliptic cohomology and modular forms, these Proceedings.

[L2] P. S. Landweber, Homological properties of comodules over $MU_*(MU)$ and $BP_*(BP)$, *Amer. J. Math.* 98 (1976), 591-610.

[LRS] P. S. Landweber, D. C. Ravenel and R. E. Stong, Periodic cohomology theories defined by elliptic curves, to appear.

[LT] J. Lubin and J. Tate, Formal complex multiplication in local fields, *Ann. of Math.* 81 (9165), 380-387.

[M] S. A. Mitchell, Finite complexes with $A(n)$-free cohomology, *Topology* 24(1985), 227-248.

[O] S. Ochanine, Sur les genres multiplicatifs définis par des intégrales elliptiques, to appear in *Topology*.

[Q] D. G. Quillen, On the formal group laws of oriented and unoriented cobordism theory, *Bull. Amer. Math. Soc., 75 (1969), 1293-1298.*

[R1] D. C. Ravenel, *Complex Cobordism and Stable Homotopy Groups of Spheres*, Academic Press, 1986.

[R2] D. C. Ravenel, Localization with respect to certain periodic homology theories, *Amer. J. Math.* 106(1984), 351-414.

[R3] D. C. Ravenel, Localization and periodicity in stable homotopy theory, to appear in Proceedings of the Durham Symposium on Homotopy Theory 1985, Cambridge University Press, 1987.

[R4] D. C. Ravenel, Formal A-modules and the Adams-Novikov spectral sequence, *J. of Pure and Applied Algebra* 32(1984), 327-345.

[Si] J. H. Silverman, *The Arithmetic of Elliptic Curves*, Springer-Verlag, 1986.

[S] E. H. Spanier, *Algebraic Topology*, McGraw-Hill, 1966.

[St] R. E. Stong, *Notes on Cobordism Theory*, Princeton Univ. Press, 1968.

[Wi] E. Witten, Elliptic genera and quantum field theory, to appear in *Math. Comm. Phys.*

DIRICHLET SERIES AND HOMOLOGY THEORY

Larry Smith
Mathematisches Institut
der Georg-August-Universität

and

R. E. Stong
Department of Mathematics
University of Virginia

§1. Introduction

Let Ω_*^U denote the complex bordism ring, and consider a ring homomorphism $\varphi: \Omega_*^U \longrightarrow Z$ into the integers. In order to avoid the trivial zero homomorphism, it will always be assumed that φ is unital, i.e., $\varphi(1) = 1$. There is then a rather obvious question.

Question. *When is φ given by an integral cohomology characteristic class?*

To clarify that question a bit, one may consider the composite $i\circ\varphi: \Omega_*^U \longrightarrow Q$ into the rational numbers. According to Hirzebruch's theory of genera [2], this genus $i\circ\varphi$ is associated with a power series

$$f(x) = \frac{x}{g^{-1}(x)} = 1+\alpha_1 x+\alpha_2 x^2+\cdots, \qquad \alpha_i \in Q,$$

and if the Chern class of a manifold M is formally expressed as $c(M) = \prod_i(1+x_i)$ via the splitting principle, then $i\circ\varphi(M) = \prod_i f(x_i)[M]$ is the value of the cohomology class

$$\prod_i f(x_i)$$

on the fundamental homology class of M. The class $\prod_i f(x_i)$ can be expressed as a linear combination of Chern classes $c_{i_1}c_{i_2}\cdots c_{i_r}$, with rational coefficients and in order that this cohomology class be an integral combination, one needs $\alpha_i \in Z$ for all i.

It is surprisingly easy to answer this question, and one has

Answer. *$\varphi: \Omega_*^U \longrightarrow Z$ is given by an integral cohomology characteristic class if and only if $n+1$ divides $\varphi(CP^n)$ for all n.*

__Proof__. Let $\tau_*: \Omega_*^U \longrightarrow H_*(BU;Z)$ be the homomorphism sending M to $\tau_*[M]$, where $\tau: M \longrightarrow BU$ classifies the tangent bundle of M. One may then consider

$$\tau_*\Omega_*^U \subset H_*(BU;Z) \subset H_*(BU;Q).$$

According to Miščenko [3], the classes $\tau_* CP^n/(n+1)$ lie in $H_*(BU;Z)$. Since $s_n[CP^n] = n+1$, one sees that $H_*(BU;Z) = Z[\tau_*(CP^n)/(n+1)]$. Thus having $\varphi(CP^n)$ divisible by n+1 for all n is equivalent to having φ extend to a ring homomorphism from $H_*(BU;Z)$ into Z. The ring homomorphisms from $H_*(BU;Q)$ into Q are given by classes $c \in H^{**}(BU;Q)$ which are grouplike, i.e., under the Whitney sum map $BU\times BU \longrightarrow BU$ one has $\Delta(c) = c \otimes c$, with Hirzebruch's series f(x) being the pullback for $CP^\infty = BU_1 \longrightarrow BU$, and the ring homomorphisms from $H_*(BU;Z)$ into Z are given exactly by those classes $c \in H^{**}(BU;Z)$. ∎

Because this question is so easy to answer, one is tempted to ask

__Question__. __When is__ φ __given by a K-theory characteristic class__?

In terms given by Hirzebruch's theory, this asks that

$$f(x) = \frac{x}{1-e^{-x}}\,(1+\alpha_1(1-e^{-x})+\alpha_2(1-e^{-x})^2+\cdots), \qquad \alpha_i \in Q,$$

should have all of the coefficients $\alpha_i \in Z$. Here the value $\varphi(M)$ is expressed as a rational linear combination of K-theory Chern numbers of M and having all α_i in Z makes this an integral linear combination.

This seems to be much more difficult but, with a bit of calculation and lots of luck, one can find an answer.

__Answer__. $\varphi: \Omega_*^U \longrightarrow Z$ __is given by a K-theory characteristic class if and only if for all primes__ p __and all integers__ r __one has__

$$\varphi(CP^{p^r-1}) \equiv \varphi(CP^{p^{r-1}-1}) \bmod p^r.$$

While split up into the different primes, this is obviously analogous to the answer for the integral cohomology and seems quite satisfactory in that respect. However, one knows that modulo decomposables CP^{p^r-1} is divisible by p^{r-1} in Ω_*^U. Thus, one realizes that these congruences modulo p^r are really mod p conditions in a disguised form.

To make this precise, one recalls the <u>Hazewinkel generators</u>

$$v_n \in \Omega_*^U \otimes Z_{(p)},$$

where $Z_{(p)} = \{\alpha/\beta \in Q \mid (\beta,p) = 1\}$ is the localization of the integers at p, which are defined by

$$CP^{p^n-1} = \sum_{i=0}^{n-1} p^{n-i-1} v_{n-i}^{p^i} CP^{p^i-1}.$$

The elements v_n are suitable generators (in their dimensions) for $\Omega_*^U \otimes Z_{(p)}$ and the Brown-Peterson homology is $BP_* = Z_{(p)}[v_n]$, which is a direct summand.

The K-theory answer just given may then be expressed as

<u>Answer′</u>. $\varphi: \Omega_*^U \longrightarrow Z$ <u>is given by a K-theory class if and only if for all primes</u> p <u>and all integers</u> r <u>one has</u>

$$\varphi(v_r) \equiv \begin{cases} 1 & r = 1 \\ 0 & r > 1 \end{cases} \mod p$$

In an analogous way, one also has

<u>Answer′</u>. $\varphi: \Omega_*^U \longrightarrow Z$ <u>is given by an integral cohomology class if and only if for each prime</u> p <u>and integer</u> r <u>one has</u>

$$\varphi(v_r) \equiv 0 \mod p.$$

At this point, one probably begins to feel that something quite general is happening. This is, of course, the case, and the goal of this paper is to describe the general phenomena.

We are indebted to the Center for Advanced Studies of the University of Virginia for providing the opportunity for us to get together. Professor Stong is indebted to the National Science Foundation for financial support during this work.

§2. The General Result

Rather than drag things out, we will begin by stating an all-encompassing result. Many of the words will undoubtedly be meaningless, but we will gradually build up the explanations in the following discussion.

Theorem. *Let $\varphi, \varepsilon : \Omega_*^U \to Z_{(p)}$ be two unital ring homomorphisms. Then the following are equivalent:*

(1) *φ and ε are (strictly) isomorphic formal groups,*

(2) *φ is given by a cohomology class in ε-theory,*

(3) *for each r, φ satisfies a congruence*
$$\varphi(CP^{p^r-1})-b_1\varphi(CP^{p^{r-1}-1})-\cdots-b_r(CP^{p^0-1}) \equiv 0 \bmod p^r,$$

(4) *for each r, φ satisfies a congruence*
$$\varphi(v_r) \equiv \varepsilon(v_r) \mod p,$$

(4′) *the induced homomorphisms*
$$\varphi|_{BP_*}, \varepsilon|_{BP_*} : BP_* \to Z/pZ$$
coincide,

(5) *the Dirichlet series of φ and ε have the same (a common) Euler factor,*

(6) *φ and ε lie in the same orbit of the action of the group G of unital ring homomorphisms from $H_*(BU;Z)$ to $Z_{(p)}$ on the set S of unital ring homomorphisms from Ω_*^U to $Z_{(p)}$.*

For the proof of the theorem, it is only necessary to read Hazewinkel's book [1]. That is a nontrivial matter, but everything needed is really standard material from formal group theory.

As the next point, in discussing integral cohomology and K-theory, the homomorphism $\varepsilon : \Omega_*^U \to Z$ was not apparent. For integral cohomology, one takes ε to be the usual augmentation for which ε sends Ω_0^U isomorphically to Z and sends all positive dimensional classes to zero. For K-theory, ε is the Todd genus homomorphism Td: $\Omega_*^U \to Z$ for which $Td(CP^n) = 1$ for all n.

As another part of the transition, we note that a ring homomorphism into Z is just a ring homomorphism into Q which goes into $Z_{(p)}$ for all p. Thus, this $Z_{(p)}$ result contains within it everything needed to understand the Z case.

Additionally, assertion number (2) is only meaningful if ε is the coefficient homomorphism of a homology theory. If ε is not known to have this property, you may suppress this statement or treat it as a vague analogy.

To begin the discussion of the result, one recalls that Quillen [5] proved that the complex bordism ring Ω_*^U is the universal formal group. Thus a ring homomorphism $\varphi\colon \Omega_*^U \longrightarrow Z_{(p)}$ is nothing more than a formal group structure for $Z_{(p)}$. Over the rational numbers, any two formal groups are equivalent, and if one considers $i\circ\varphi\colon \Omega_*^U \longrightarrow Q$, the Hirzebruch theory of genera contains the relevant information about the formal group. Specifically, the series associated with $i\circ\varphi$

$$f(x) = \frac{x}{g^{-1}(x)} = 1+\alpha_1 x+\alpha_2 x+\alpha_2 x^2+\cdots, \qquad \alpha_i \in Q,$$

gives the formula for the change of power series variable. One lets

$$u = g^{-1}(x) = x/(1+\alpha_1 x+\alpha_2 x^2+\cdots),$$

$$= x+\beta_1 x+\beta_2 x^2+\cdots, \qquad \beta_i \in Q,$$

and the formal group law corresponding to φ is given by

$$g^{-1}(g(u)+g(v)) \in Z_{(p)}[[u,v,]].$$

If one now considers a second ring homomorphism $\varepsilon\colon \Omega_*^U \longrightarrow Z_{(p)}$ for which $i\circ\varepsilon\colon \Omega_*^U \longrightarrow Q$ has the Hirzebruch series

$$f_\varepsilon(x) = \frac{x}{g_\varepsilon^{-1}(x)},$$

then one may write

$$\frac{x}{g_\varphi^{-1}(x)} = f_\varphi(x) = \frac{x}{g_\varepsilon^{-1}(x)}\,(1+\gamma_1 g_\varepsilon^{-1}(x)+\gamma_2 g_\varepsilon^{-1}(x)^2+\cdots), \quad \gamma_i \in Q$$

so that

$$g_\varphi^{-1}(x) = g_\varepsilon^{-1}(x)/(1+\gamma_1 g_\varepsilon^{-1}(x)+\gamma_2 g_\varepsilon^{-1}(x)^2+\cdots),$$

$$= g_\varepsilon^{-1}(x)+\delta_2 g_\varepsilon^{-1}(x)^2+\delta_3 g_\varepsilon^{-1}(x)^3+\cdots,$$

with the coefficients $\delta_i \in Q$. Of course, this formula gives the isomorphism of the rational formal groups $i\circ\varphi$ and $i\circ\varepsilon$.

In order that φ and ε be isomorphic formal groups over $Z_{(p)}$, one needs the coefficients δ_i to belong to $Z_{(p)}$. By the formula for inversion of series of the form 1+y, this is equivalent to having $\gamma_i \in Z_{(p)}$ for all i. This condition is, in turn, precisely equivalent to saying that $\varphi(M)$ can be expresssed in terms of $Z_{(p)}$ linear combinations of the Chern classes in ε-theory, via the formula for $f_\varphi(x)$ in terms of $g_\varepsilon^{-1}(x)$.

This shows that conditions (1) and (2) are equivalent.

We now need to recall some things about formal groups, so let's begin with a ring homomorphism $\varphi\colon \Omega_*^U \longrightarrow Q$. One can then form a power series

$$\hat{\varphi}(z) = \sum_{n\geq 0} \varphi(CP^n)z^n = 1+\sum_{n>0}\varphi(CP^n)z^n \in Q[[z]]$$

and

$$x = g_\varphi(u) = \int_0^u \hat{\varphi}(z)dz = \sum_{n\geq 0} \frac{\varphi(CP^n)}{n+1} z^{n+1}$$

is the logarithm of the formal group of φ given by

$$g_\varphi^{-1}(g_\varphi(u)+g_\varphi(v)),$$

as above.

Additionally, one can associate to φ a <u>Dirichlet series</u>

$$L_\varphi(s) = \sum_{n\geq 1} \varphi(CP^{n-1})n^{-s}.$$

For example, if φ is the Todd homomorphism, with $\varphi(CP^n) = 1$ for all n, then $\hat{\varphi}(z) = \Sigma z^n = 1/(1-z)$ and $L_\varphi(x) = \sum_{n\geq 1} n^{-s}$ is the classical zeta function. The nice result, making this more than a formal change of perspective is

<u>Theorem</u> (Hazewinkel [1;§33]). <u>If $\varphi\colon \Omega_*^U \longrightarrow Q$ is a unital ring homomorphism sending the subring $Z[CP^n]$ into $Z_{(p)}$, then</u>

$$\varphi\colon \Omega_*^U \longrightarrow Z_{(p)}$$

<u>if and only if</u> $L_\varphi(s)$ <u>has an Euler factor</u> (<u>at the prime</u> p), <u>i.e.</u>, <u>there is a factorization</u> (<u>in</u> $\mathbf{Z}_p$ = p-<u>adics</u>)

$$L_\varphi(s) = (1-b_1p^{-s}-b_2p^{-2s}-\cdots)^{-1}(\sum_{m\geq 1} c_m m^{-s})$$

<u>where</u> $c_{p^r q}$ <u>is divisible by</u> p^r <u>and</u> p^{i-1} <u>divides</u> b_i.

<u>Note</u>: While Hazewinkel says factorization is in $\mathbf{Z}_p$, the factorization is purely formal.

At first glance, this looks fairly mysterious, but what it is really saying is that the p-primary divisibility which takes place in Ω_*^U can be nicely described in terms of the Dirichlet series.

Notice that <u>the first factor</u>

$$(1-b_1p^{-s}-b_2p^{-2s}-\cdots)^{-1}$$

is the term which is called the Euler factor. For the classical Riemann zeta function $\sum_{n\geq 1} n^{-s} = \prod_p (1-p^{-s})^{-1}$, the Euler factor (at p) is $(1-p^{-s})^{-1}$.

For our purposes, it is crucial to ask, "<u>How unique is the Euler factor</u>?" While this is undoubtedly well known, let's consider

$$(1-b_1p^{-s}-b_2p^{-2s}-\cdots)^{-1}(\Sigma\, c_m m^{-s}) = (1-b_1'p^{-s}-b_2'p^{-2s}-\cdots)^{-1}(\Sigma\, c_m' m^{-s})$$

and note that we have $c_1 = c_1' = 1$ (since $\varphi(CP^0) = 1$). Now the product of two Dirichlet series with lead term 1

$$(\Sigma\, c_m m^{-s})(\Sigma\, d_m m^{-s}) = (\Sigma\, e_m m^{-s})$$

where

$$e_n = \sum_{j|n} c_j d_{n/j} = c_n + d_n + \sum_{\substack{j|n \\ 1<j<n}} c_j d_{n/j}$$

shows that you can inductively invert series with leading term 1. If we let $\nu_p(n)$ be the power of p dividing n, then $\nu_p(c_j) \geq \nu_p(j)$ and $\nu_p(d_{n/j}) \geq \nu_p(n/j)$ gives $\nu_p(c_j d_{n/j}) \geq \nu_p(c_j)+\nu_p(d_{n/j}) \geq \nu_p(j)+\nu_p(n/j) = \nu_p(n)$, so for the product, $\nu_p(e_n) \geq \nu_p(n)$.

From the presumed equality we have

$$(1-b_1p^{-s}-b_2p^{-2s}-\cdots)^{-1}(1-b_1'p^{-s}-\cdots) = (\Sigma\, c_m m^{-s})^{-1}(\Sigma\, c_m' m^{-s}),$$

$$= \Sigma\, \alpha_m m^{-s}$$

where from the right hand side of the equation we have α_{p^rq} divisible by p^r and from the left hand side

$$\Sigma\, \alpha_m m^{-s} = \sum_r \alpha_{p^r}\, p^{-rs}$$

and α_{p^r} is divisible by p^r. Thus, if $(1-b_1p^{-s}-b_2p^{-2s}-\cdots)^{-1}$ is one Euler factor (recall that p^{i-1} divides b_i), then any other has the form

$$((1-b_1p^{-s}-b_2p^{-2s}-\cdots)(\Sigma\, \alpha_{p^r}\, p^{-rs}))^{-1}$$

where p^r divides α_{p^r}.

Now let's look at

$$L_\varphi(s) = \Sigma\, \varphi(CP^{n-1})n^{-s}$$

$$= ((1-b_1p^{-s}-b_2p^{-2s}-\cdots)(\Sigma\, \alpha_{p^r}\, p^{-rs}))^{-1}((\Sigma\, \alpha_{p^r}\, p^{-rs})(\Sigma c_m m^{-s}))$$

and <u>choose</u> $\Sigma\, \alpha_{p^r}\, p^{-rs}$ to be the series $(\Sigma\, c_{p^r}\, p^{-rs})^{-1}$. This gives

$$L_\varphi(s) = (1-b_1'p^{-s}-\cdots)(\Sigma\, c_m' m^{-s})$$

where $c'_{p^rq} \equiv 0 \bmod p^r$, <u>and</u> $c'_{p^r} = 0$, i.e., if $q = 1$, you have zero. Then

$$\sum_r \varphi(CP^{p^r-1})p^{-rs} = (1-b_1'p^{-s}-b_2'p^{-2s}-\cdots)^{-1}$$

and we have

<u>**Observation**</u>. <u>One choice for the Euler factor</u> (if it exists) <u>is to take</u>

$$1-b_1p^{-s}-b_2p^{-2s}-\cdots = (\sum_r \varphi(CP^{p^r-1})p^{-rs})^{-1}.$$

<u>Note</u>: If $\varphi: \Omega_*^U \longrightarrow Q$ with $\varphi(CP^n) \in Z_{(p)}$ for all n, then one can write

$$L_\varphi(s) = \Sigma\, \varphi(CP^{n-1})n^{-s} = (\sum_r \varphi(CP^{p^r-1})p^{-rs})(\Sigma\, d_m m^{-s})$$

where $d_{p^r} = 0$. <u>In order that</u> $\varphi: \Omega_*^U \longrightarrow Z_{(p)}$, <u>it is necessary and sufficient that</u>

(1) $(1-b_1p^{-s}-b_2p^{-2s}-\cdots) = (\sum_r \varphi(CP^{p^r-1})p^{-rs})^{-1}$ <u>satisfy</u> p^{i-1} <u>divides</u> b_i and

(2) $\varphi(CP^{p^rq-1})-b_1\varphi(CP^{p^{r-1}q-1})-\cdots-b_r\varphi(CP^{q-1}) \equiv 0 \pmod{p^r}$ <u>for all</u> <u>$q > 1$</u>. (For $q = 1$, this expression is zero and defines b_i).

This describes the p-primary divisibilities that are satisfied among the CP^n in Ω_*^U.

Having digressed a bit, we can now return to our main result and see that $\varphi, \varepsilon: \Omega_*^U \longrightarrow Z_{(p)}$ have the same (or a common) Euler factor if and only if

$$\varphi(CP^{p^r-1}) \equiv b_1\varphi(CP^{p^{r-1}-1})+\cdots+b_r\varphi(CP^{p^0-1}) \bmod p^r$$

where the b_i are defined by

$$\varepsilon(CP^{p^r-1}) = b_1\varepsilon(CP^{p^{r-1}-1})+\cdots+b_r\varepsilon(CP^{p^0-1})$$

(or, if you prefer, the b_i are given by some other Euler factor for $L_\varepsilon(s)$). This describes the undefined b_i in condition (3) and establishes the equivalence of conditions (3) and (5).

<u>Note</u>: For integral cohomology, $L_\varepsilon(s) = 1$ and $b_i = 0$ for all i. Thus, the congruences become $\varphi(CP^{p^r-1}) \equiv 0$. Further, the fact that φ maps into $Z_{(p)}$ automatically gives $\varphi(CP^{p^rq-1}) \equiv 0$ for all $q > 1$. For K-theory, $L_\varepsilon(s) = \Sigma\, n^{-s}$ had Euler factor $(1-p^{-s})^{-1}$, so $b_1 = 1$ and $b_i = 0$ for $i > 1$. The general result reduces to the congruence $\varphi(CP^{p^r-1}) \equiv \varphi(CP^{p^{r-1}-1}) \bmod p^r$.

As a next step, let's establish the equivalence of (5) and (6), So we let G be the set of unital ring homomorphisms from $H_*(BU;Z)$ to

$Z_{(p)}$, and S the set of unital ring homomorphisms from Ω_*^U into $Z_{(p)}$. Further, we let T be the set of unital ring homomorphisms from Ω_*^U into Q which send $Z[CP^n]$ into $Z_{(p)}$. One then has a bijection

$$T \longrightarrow \begin{cases} \text{Dirichlet series with } Z_{(p)} \text{ coefficients having} \\ \text{leading term } 1 \end{cases}$$

by sending φ to the series $L_\varphi(s) = \Sigma\, \varphi(CP^{n-1})n^{-s}$. Using the multiplication of Dirichlet series, this gives T the structure of an abelian group (series beginning with 1 were invertible).

Inside T one has a subgroup given by G, which is the group of series $\Sigma\, \varphi(CP^{n-1})n^{-s}$ for which $\varphi(CP^{p^r q-1}) \equiv 0 \bmod p^r$. (Notice: these are the series which make up the non-Euler factors.) It is then clear that two series L_φ and L_ε with $\varphi, \varepsilon \in S \subset T$ have the same Euler factor if and only if they lie in the same orbit of G.

Thus, condition (6) just recognizes the group structure implicit in the product of Dirichlet series.

In order to prove the equivalence of (4) and (5), we consider $\varphi, \varepsilon : \Omega_*^U \longrightarrow Z_{(p)}$ and write

$$\Sigma\, \varphi(CP^{p^r-1})p^{-rs} = (\Sigma\, \varepsilon(CP^{p^r-1})p^{-rs})(\Sigma\, \alpha_r p^{-rs})$$

assuming that $\alpha_k \equiv 0 \bmod p^k$ for $k < r$ and $\varphi(v_i) \equiv \varepsilon(v_i) \bmod p$ for $i < r$. We want to prove that $\alpha_r \equiv 0 \bmod p^r$ if and only if $\varphi(v_r) \equiv \varepsilon(v_r) \bmod p$.

To begin

$$\alpha_r = \varphi(CP^{p^r-1}) - \sum_{s<r} \varepsilon(CP^{p^{r-s}-1})\alpha_s$$

$$= \sum_{i<r} p^{r-1-i}\varphi(v_{r-i})^{p^i}\varphi(CP^{p^i-1}) - \sum_{s<r} \varepsilon(CP^{p^{r-s}-1})\alpha_s$$

and

$$\alpha_r - p^{r-1}\varphi(v_r) = \sum_{0<i<r} p^{r-1-i}\varphi(v_{r-i})^{p^i}\varphi(CP^{p^i-1}) - \sum_{s<r} \varepsilon(CP^{p^{r-s}-1})\alpha_s.$$

Now $0 < i < r$ gives $\varphi(v_{r-i})^{p^i} \equiv \varepsilon(v_{r-i}) \bmod p$ and so $\varphi(v_{r-i})p^i \equiv \varepsilon(v_{r-i})p^i \bmod p^{i+1}$ or

$$\alpha_r - p^{r-1}\varphi(v_r)$$

$$\equiv \sum_{0<i<r} p^{r-1-i}\varepsilon(v_{r-i})^{p^i}\varphi(CP^{p^i-1}) - \sum_{s<r}\varepsilon(CP^{p^{r-s}-1})\alpha_s.$$

$$\equiv \sum_{0<i<r} p^{r-1-i}\varepsilon(v_{r-i})^{p^i}\sum_{0\leq t\leq i}\varepsilon(CP^{p^{i-t}-1}) - \sum_{s<r}\varepsilon(CP^{p^{r-s}-1})\alpha_s.$$

mod p^r. Now the sum with $0 < i < r$ and $0 \leq t \leq i$ is the same as the sum with $0 \leq t < r$ and $t \leq i < r$, except for the term $t = i = 0$, so

$$\alpha_r - p^{r-1}\varphi(v_r) + p^{r-1}\varepsilon(v_r)$$

$$= \sum_{t<r}\alpha_t\left(\sum_{t\leq i<r} p^{r-1-i}\varepsilon(v_{r-i})^{p^i}\varepsilon(CP^{p^{i-t}-1})\right) - \sum_{s<r}\varepsilon(CP^{p^{r-s}-1})\alpha_s.$$

$$= \sum_{s<r}\alpha_s\left\{\sum_{s\leq i<r} p^{r-s-1-(i-s)}\varepsilon(v_{r-s-(i-s)})^{p^{s+(i-s)}}\varepsilon(CP^{p^{i-s}-1}) - \varepsilon(CP^{p^{r-s}-1})\right\}$$

(putting $t = s$)

$$= \sum_{s<r}\left\{\sum_{j<r-s} p^{r-s-1-j}\varepsilon(v_{r-s-j})^{p^{s+j}}\varepsilon(CP^{p^j-1}) - \varepsilon(CP^{p^{r-s}-1})\right\}$$

(putting $j = i-s$). Now $\varepsilon(v_{r-s-j})^{p^s} \equiv \varepsilon(v_{r-s-j})$ mod p so $\varepsilon(v_{r-s-j})^{p^{s+j}} \equiv \varepsilon(v_{r-s-j})^{p^j}$ mod p^{j+1} and multiplying by $p^{r-s-1-j}\alpha_s$, with $\alpha_s \equiv 0$ mod p^s since $s < r$ gives

$$\alpha_s p^{r-s-1-j}\varepsilon(v_{r-s-j})^{p^{s+j}} \equiv \alpha_s p^{r-s-1-j}\varepsilon(v_{r-s-j})^{p^j} \text{ mod } p^r.$$

Thus

$$\alpha_r - p^{r-1}\varphi(v_r) + p^{r-1}\varepsilon(v_r) \equiv \sum_{s<r}\alpha_s\left\{\sum_{j<r-s} p^{r-s-1-j}\varepsilon(v_{r-s-j})^{p^j}\varepsilon(CP^{p^j-1}) - \varepsilon(CP^{p^{r-s}-1})\right\}$$

$$\equiv 0 \text{ mod } p^r,$$

since the coefficient of α_s is zero. This gives

$$\alpha_r \equiv p^{r-1}(\varphi(v_r)-\varepsilon(v_r)) \mod p^r$$

and $\alpha_r \equiv 0 \mod p^r$ if and only if $\varphi(v_r) \equiv \varepsilon(v_r) \mod p^r$.

Then (4) and (4′) are obviously equivalent since $BP_* = Z_{(p)}[v_n]$.

Finally, the equivalence of (1) and (5) can be established by direct calculation. It is not easy and certainly isn't pleasant. To avoid that one can refer to Hazewinkel [1]. Equation (33.15) says that the logarithm series satisfies a functional equation given entirely in terms of the Euler factor. Then §2.2 says that two formal group laws are strictly isomorphic if and only if they satisfy functional equations of the same type. Thus the equivalence we want is a consequence of the functional equation.

§3. Some Sidelights

There are several points which are worthy of some comment.

Comment 1. The defining relation

$$CP^{p^{n-1}} = \sum_{i=0}^{n-1} p^{n-i-1} v_{n-i}^{p^i} CP^{p^i-1}$$

says that the series formed using v's is almost the inverse of the CP series, hence should lead to an Euler factor. Explicitly, one may let $\varepsilon : \Omega_*^U \longrightarrow Z_{(p)}$ and form

$$\widetilde{CP} = 1+\varepsilon(CP^{p-1})p^{-s}+\varepsilon(CP^{p^2-1})p^{-2s}+\cdots$$

and

$$\tilde{\varepsilon}_{(r)} = 1-p^0\varepsilon(v_1)^{p^{r-1}}p^{-s}-\cdots-p^{r-1-i}\varepsilon(v_{r-i})^{p^i}p^{-(r-i)s}-\cdots$$

$$-p^{r-1}\varepsilon(v_r)^{p^0}p^{-rs}-\sum_{t>r} p^{t-1}\varepsilon(v_t)^{p^0}p^{-ts}.$$

In the product $\tilde{\varepsilon}_{(r)}\widetilde{CP}$ the coefficient of p^{-ks}, with $k \leq r$, is

$$\varepsilon(CP^{p^k-1}) - \sum_{i<k} p^{k-i-1}\varepsilon(v_{k-i})^{p^{r-k+i}}\varepsilon(CP^{p^i-1})$$

which is zero for $k = r$. For $k < r$, $\varepsilon(v_{k-i})^{p^{r-k}} \equiv \varepsilon(v_{k-i}) \mod p$, so $\varepsilon(v_{k-i})^{p^{r-k+i}} \equiv \varepsilon(v_{k-i})^{p^i} \mod p^{i+1}$ and the coefficient is congruent to

$$\varepsilon(CP^{p^k-1}) - \sum_{i<k} p^{k-i-1}\varepsilon(v_{k-i})^{p^i}\varepsilon(CP^{p^i-1}) = 0 \bmod p^k.$$

By noting that $\varepsilon(v_j)^{p^{s-1}}$ is well defined in Z/p^rZ for $s \geq r$, if $\varepsilon(v_j)$ is known in Z/pZ one may pass to the limit on r to obtain

__Proposition__. __For__ $\Sigma\, \varepsilon(CP^{p^r-1})p^{-rs} = (1-b_1p^{-s}-b_2p^{-2s}-\dots)^{-1}$ __one has a factorization__

$$1-b_1p^{-s}-b_2p^{-2s}-\dots = (1-p^0\varepsilon(v_1)^{p^\infty}p^{-s}-\dots-p^{k-1}\varepsilon(v_k)^{p^\infty}p^{-ks}-\dots)$$
$$\cdot\ (\sum_{t\geq 0}\beta_t p^{-ts})$$

__in__ $\mathbf{Z}_p$ (__the__ __p-adics__) __where__

$$\varepsilon(v_i)^{p^\infty} = \lim_{r\to\infty}\varepsilon(v_i)^{p^r}$$

__depends__ __only__ __on__ $\varepsilon(v_i)$ __in__ Z/pZ __and__ β_t __is__ __divisible__ __by__ p^t.

If one has $\varphi,\varepsilon: \Omega_*^U \longrightarrow Z_{(p)}$ with $\varphi(v_i) \equiv \varepsilon(v_i) \bmod p$, then

$$\Sigma\, \varphi(CP^{p^r-1})p^{-rs} = (\Sigma\, \varepsilon(CP^{p^r-1})p^{-rs})(\Sigma\, \gamma_t p^{-ts})$$

where __first__, one has $\gamma_t \in Z_{(p)}$ just by division, and __second__, one has $\gamma_t \equiv 0 \bmod p^t$ in $\mathbf{Z}_p$ (the p-adics). Thus the two homomorphisms have the same Euler factor.

__Comment 2__. From the equivalence of (1), (6), and (4′), one sees that the set of strict isomorphism classes of formal group laws $\varphi: \Omega_*^U \longrightarrow Z_{(p)}$ or equivalently the orbit space S/G can be identified as a subset of $RHom(BP_*, Z/pZ)$, the ring homomorphisms from $BP_* = Z_{(p)}[v_i]$ into Z/pZ.

To see that you actually get all ring homomorphisms, one need only define a homomorphism $\varphi: \Omega_*^U \longrightarrow Z_{(p)}$ by letting $L_\varphi(s) = (1-b_1p^{-s}-b_2p^{-2s}-\dots)^{-1}$ be just an Euler factor where $b_i = p^{i-1}b_i'$,

$b_i' \in Z_{(p)}$ being chosen arbitrarily. Then the b_i' in Z/pZ are precisely the values $\varphi(v_i)$ in Z/pZ.

Jack Morava [4] considered several particularly interesting ring homomorphisms formed in exactly this way.

Comment 3. One would love to be able to do this same sort of thing with homomorphisms $\varphi: \Omega_*^U \longrightarrow A$ where A was some ring other than Z or $Z_{(p)}$ and, in particular, the universal example with $A = \Omega_*^U$ would be exciting. However, it really can't be done. The arguments make crucial use of the fact that $x^p \equiv x \bmod p$ holds in A.

Comment 4. As a curiosity, one can consider those ring homomorphisms $\varphi: \Omega_*^U \longrightarrow Z$ for which $\varphi(CP^n)$ is either zero or one for each n. The three obvious examples are

(1) the augmentation for ordinary homology,

(2) the Todd genus, and

(3) the Hirzebruch signature.

From the relations

$$\varphi(CP^{p^r q-1})-b_1\varphi(CP^{p^{r-1}q-1})-\cdots-b_r\varphi(CP^{q-1})\begin{cases} \equiv 0 \ (p^r) \text{ if } q > 1 \\ = 0 \text{ if } q = 1 \end{cases}$$

where $p^{i-1} | b_i$, it follows that

$$\varphi(CP^{n-1}) = \prod_{p|n} \varphi(CP^{p-1})$$

and that the choice of the values $\varphi(CP^{p-1})$ with p prime can be made arbitrarily. For each prime p one can find either a K-theory or ordinary cohomology Chern class giving rise to φ into $Z_{(p)}$. Hence, one has maps of ring spectra

$$\begin{array}{ccc} MU & \longrightarrow & BU \otimes Z_{(S)} \\ \downarrow & & \downarrow \\ K(Z[t,t^{-1}]) \otimes Z_{(T)} & \longrightarrow & K(Q[t,t^{-1}]) \end{array}$$

into the localization of the K-theory spectrum for one set of primes (S) and of homology at a set (T) with $T \cap S = \emptyset$, $T \cup S = \{\text{primes}\}$, agreeing when pushed into the Eilenberg-MacLane spectrum $K(Q[t,t^{-1}])$. $\varphi: \Omega_*^U \longrightarrow Z$ is then realized by mapping into the ring spectrum which is the fiber product, and is the orientation homomorphism of a homology theory which is K-theory at one set of primes and ordinary homology at the others.

__Note__: For the Hirzebruch signature, the L polynomials have odd denominators which is giving ordinary homology over $Z_{(2)}$. One also knows that $\Omega_*^U(\ ) \otimes_L Z[1/2] = KO_*(\ ) \otimes Z[1/2]$, which is a direct summand of $K_*(\ ) \otimes Z[1/2]$. This case is then well known.

This illustrates the point that if $\varepsilon: \Omega_*^U \longrightarrow Z_{(p)}$ is realized by a map of ring spectra $MU \longrightarrow E_\varepsilon$ and if $\varphi: \Omega_*^U \longrightarrow Z_{(p)}$ is strictly isomorphic to ε, then φ is also realized by such a map (given by changing the map into E_ε by multiplying by an appropriate Chern class in ε-theory).

In fact, it is sufficient to realize ε by a map $MU \xrightarrow{\varepsilon} E_\varepsilon$ of MU module spectra for one may use a Chern class in MU theory, or more precisely, $MUZ_{(p)}$. One need only consider the composite

$$MU \longrightarrow S^0 \wedge MU \xrightarrow{\text{unit}\wedge 1} MU \wedge MU \xrightarrow{c\wedge 1} MUZ_{(p)} \wedge MUZ_{(p)} \longrightarrow$$

$$\xrightarrow{\varepsilon\wedge 1} E_\varepsilon \wedge MUZ_{(p)} \longrightarrow E_\varepsilon$$

using the local property of E_ε, where $c \in \tilde{\Omega}_*^U(MU) \otimes Z_{(p)}$ is any grouplike element.

It also illustrates the point that one may assemble homology theories for different primes by using the fiber product spectrum. (Note: It seems convenient to use periodic theories mapping into $K(Q[t,t^{-1}])$ with degree $t = 2$, so that $\varphi: \Omega_*^U \longrightarrow Z_{(p)}$ is replaced by $\Phi: \Omega_*^U \longrightarrow Z_{(p)}[t,t^{-1}]$ with $\Phi(M^{2n}) = \varphi(M)t^n$.)

__References__

1. Hazewinkel, Michiel: __Formal Groups and Applications__. Academic Press, New York, 1978.

2. Hirzebruch, F.: __Topological Methods in Algebraic Geometry__. Springer-Verlag, Berlin, 3rd ed., 1966.

3. Miščenko, A. S.: Appendix 1 in S. P. Novikov: The methods of algebraic topology from the viewpoint of cobordism theory, Izv. Akad. Nauk. SSSR, Ser. Mat. Tom 31 (1967) no. 4, 855-956; Math USSR-Izvestija 1 (1967), 827-913.

4. Morava, Jack: Forms of K-theory, unpublished manuscript.

5. Quillen, Daniel: On the formal group laws of unoriented and complex cobordism theory, Bull. Amer. Math. Soc. 75 (1969), 1293-1298.

CONSTRAINED HAMILTONIANS
An Introduction to Homological Algebra in Field Theoretical Physics

James D. Stasheff
Department of Mathematics, University of North Carolina
Chapel Hill, North Carolina 27514

Cohomological physics has been recognized only in the last decade or two, but implicitly goes back much further, though not as far back as the idea of a field. In physical terms, that idea can be said to go back at least to 1269, when Petrus Perégrinus described the magnetic field around a lodestone [17]. This is the easiest field to visualize (literally), as you all have seen, although one can also see the gravitational and electric fields. These all correspond to vector fields in the mathematical sense of tangent vector fields, the manifold being our three dimensional one.

The cohomological aspect did not appear until after Minkowski combined the electric field E and the magnetic field B in a differential 2-form:

$$F = \Sigma E_i dx^i dt + \Sigma B_i dx^j dx^k \quad \text{(cyclic sum)}.$$

Maxwell's equations can then be written succinctly as

$$dF = 0$$
$$d{*}F = 0$$

in the source free case, where * is the Hodge *-operator for Minkowski space.

The first relevance of cohomology appears in Dirac's treatment of the magnetic monopole - discovering, in physical theory, the Hopf fibration $S^1 \to S^3 \to S^2$ in 1931!

With the development of gauge field theories such as Yang-Mills, de Rham cohomology and characteristic classes became of considerable interest as did other cohomologies, especially those for smooth groups and Lie algebras and representations. Today I'd like to tell you about some developments of the past decade in which the simplicity of Lie algebra representation has given way to a more subtle structure and added homological algebra to the list of physically relevant tools. The generalization which physical intuition forced on those doing certain calculations in gravity turned out to be the same generalization that occurred quite independently in the deformation theory of algebras and rational homotopy theory, in particular, in my work with Mike Schlessinger.

To begin on the physical side, where I speak as a tourist, first there was the work of Fradkin and his school, particularly Batalin and Vilkovisky [9,1,2]. An excellent report by Henneaux [13] called attention to the homological aspect of their technique which was further elucidated by McMullan [6,16] who recognized the Koszul complex and its crucial role therein.

The problem is posed in the setting of the Hamiltonian formalism, which includes the following crucial (for us) ingredients:

A phase space W, e.g., the cotangent bundle T^*M or more generally a symplectic manifold, i.e., a smooth manifold with a closed 2-form of maximal rank ($dp_i \wedge dq_i$ in local coordinates). This gives a Poisson algebra structure on $C^\infty(W)$, i.e., a Poisson bracket $\{\ ,\ \} : C^\infty(W) \otimes C^\infty(W) \longrightarrow C^\infty(W)$ making $C^\infty(W)$ a Lie algebra over R and, in addition, satisfying

$$\{f,gh\} = \{f,g\}h + f\{g,h\} .$$

Thus $\{f,\ \}$ is a derivation of $C^\infty(W)$ and so can be identified with a vector field denoted X_f.

In field theory, W is $C^\infty(M)$ or $Sec(E\to M)$.

Now the problem that leads to today's talk is the problem of "constraints", i.e., a family of functions $\{\varphi_\alpha\}$ such that the dynamics is constrained to stay in the submanifold $V \subset W$ which is the common zero set of the φ_α. Worse yet, the submanifold V does not reflect the true "physical degrees of freedom", but rather the constraints also act on V via their associated vector fields X and this action takes flows (solutions) to physically equivalent ones. The constraints have determined the foliation $\mathcal{F}$ of V and the quotient $V/\mathcal{F}$ is the "reduced phase space" which is the physically meaningful symplectic manifold.

For example:

<u>Gauge theory</u>: Here W is $T^*\mathcal{A}$ where $\mathcal{A}$ is the space of connections for a fixed principal G-bundle $G \to P \to X$. The reduced phase space is $T^*(\mathcal{A}/\mathcal{G})$ where $\mathcal{G}$ is the group of "gauge transformations", i.e., vertical automorphisms of P, that is,

$$\mathcal{G} = Sec(P\times_{Ad}G) .$$

<u>Gravity</u>: Here W is $T^*\mathcal{M}$ where $\mathcal{M}$ is the space of metrics on a fixed manifold X. The reduced phase space is $T^*(\mathcal{M}/\mathrm{Diff}\ X)$.

String theory and more general non-linear Σ-models. Here W is at least as big as $T^*(M^\Sigma)$ for some Riemann surface Σ or other manifold.

Now, for a mathematician, there's a tendency to think that we have a perfectly adequate description of the reduced phase space, but to a physicist, since the original data are on W or even $\mathcal{A}$ and $\mathcal{M}$, there is a need for handling computations in terms of W directly - without passing to the quotient. The method physicists have developed is to add ghost fields to the situation - you can't see ghosts experimentally, but they account computationally for what you do see - and implementing a BRS(T) transformation.

What is a BRS(T) operator? First, BRS refers to Becchi-Rouet-Stora [3], while the (optional) T refers to Tyutin, whose preprint has never been published, nor have I been able to locate a copy. At first, the BRS operator appeared to be a formal construct corresponding to a symmetry of certain Lagrangians. Later, it was identified via the exterior derivative d_P on the forms on the principal G-bundle $G \to P \to M$. More precisely, it could be identified with the model of the structure of $G \to P \to M$ in terms of $\Lambda g^* \otimes \Omega^*(P)$ with the Cartan-Chevalley-Eilenberg differential D. Here g is the Lie algebra of G and in the finite dimensional case, Λg^* is the exterior or Grassmann algebra on the dual of g.

For an arbitrary g, we interpret Λg^* as $(\Lambda g)^* \approx \mathrm{Alt}(g,\mathbb{R})$ = alternating multilinear functions from g to $\mathbb{R}$. Define

$$(\delta_g h)(X_o,\ldots,X_p) = \Sigma(-1)^{i+j} h\,([X_i,X_j],\ldots\hat{i}\ldots\hat{j}\ldots)$$

where $\hat{i}$ denotes omission of X_i.

$(\delta_g)^2 = 0$ if and only if $[\,,]$ satisfies the Jacobi identity.

For a linear map $\theta : g \to \mathrm{Aut}\, M$ of a vector space M, δ_θ is defined on $\mathrm{Alt}(g,M)$ by

$$(\delta_\theta h)(X_o,\ldots,X_p) = \Sigma(-1)^i \theta(X_i)h(\ldots\hat{i}\ldots)\ .$$

If $(\delta_g)^2 = 0$, then $(\delta_g+\delta_\theta)^2 = 0$ iff θ is a representation:

$$\theta([X,Y]) = \theta(X)\theta(Y) - \theta(Y)\theta(X)\ .$$

Given all of this,

$$H_{Lie}(g,M) \text{ means } \frac{\operatorname{Ker} \delta_g + \delta_\theta}{\operatorname{Im} \delta_g + \delta_\theta} .$$

The relevance to constraints is that

$$H^{\circ}_{Lie}(g,M) = g\text{-invariants} \subset M .$$

Gradually, BRS(T) has come to refer to various generalizations of $D = \delta_g + \delta_\theta$, including the one we'll study today.

This Lie algebra cohomology is relevant to the constraint problem because, for our picture

$$\begin{array}{ccc} V & \subset & W \\ \downarrow & & \\ V/\mathcal{F} , & & \end{array}$$

we have $C^\infty(V) \approx C^\infty(W)/I$ where I is the ideal of constraints, $I = \{f^\alpha \varphi_\alpha | f^\alpha \in C^\infty(W)\}$. If $\mathcal{F}$ were given by a group action, $C(V/G)$ would be given by the G-equivariant functions on $C(V)$. For a connected Lie group G, it would be sufficient to look at the g-equivariant functions, i.e., the g-invariant elements of $C^\infty(W)/I$ which is precisely $H^{\circ}_{Lie}(G,C^\infty(W)/I)$.

Now $\mathcal{F}$ is generated by the φ_α. Do they form a Lie algebra? NO. Do they span a Lie algebra? Yes and no. I want to consider just the physically difficult case of FIRST CLASS constraints, meaning $\{\varphi_\alpha, \varphi_\beta\} = C^\gamma_{\alpha\beta}\varphi_\gamma$. If we look at the real vector space spanned by the constraints, we have a Lie algebra provided the structure functions $C^\gamma_{\alpha\beta}$ are constants, but in general they are functions. Then the real span Φ of the constraints need not be a Lie algebra, though I, the span over $A = C^\infty(W)$ of the constraints, certainly is a Lie algebra over R, but not over A since

$$\{\varphi_\alpha, f\, \varphi_\beta\} = \{\varphi_\alpha, f\}\varphi_\beta + f\{\varphi_\alpha, \varphi_\beta\} .$$

Thus we can look at the Lie algebra cohomology $H_{Lie}(I,A/I)$ and know that H°_{Lie} will give us the desired adI-invariant elements of A/I. But what of the higher H^1? And should we ignore the Poisson algebra structure, i.e., the distributivity? In fact, the A-multilinear functions

$$\mathrm{Alt}_A(I,A/I)$$

do form a subcomplex of $\mathrm{Alt}(I,A/I)$ (though this is not true for a general I-module) and give the same H°. One of our new insights into the FBV formalisms is the relevance of this subcomplex.

The goal of Fradkin et al was to describe their dynamics in terms of the constraints φ_α, i.e., in terms of their real span Φ. For this, they introduced ghosts.

What are ghosts? They are generators ρ_α of a Grassmann algebra over A in one-to-one correspondence with the constraints φ_α. In this context of FIRST CLASS constraints, I will refer to them as Koszul ghosts for the regular ideal I, for indeed the Koszul resolution $K(I)$ of A/I is the exterior algebra over A:

$$K(I) = A \otimes \Lambda\rho_\alpha$$

with $d_K : K(I) \longrightarrow K(I)$ being the A-<u>derivation</u> defined by $d_K\rho_\alpha = \varphi_\alpha$. The ghost degree of ρ_α is one and is logarithmic so that the ghost degree of a monomial is just the number of ρ_α's.

A word about notation: In the physics literature, with the exception of McMullan, the derivation d_K is written $\varphi_\alpha\eta^\alpha$ where η^α is interpreted as dual to ρ_α, i.e., $< \eta^\alpha, \rho_\rho > = \delta^\alpha{}_\beta$.

Fradkin et al, Henneaux and McMullan also consider the possibility of starting with fermions as well as bosons, i.e., starting with a graded commutative algebra A with homogeneous constraints φ_α. The ghost ρ_α then has the opposite parity to φ_α, in fact degree ρ_α = degree $\varphi_\alpha+1$.

That $K(I)$ is a resolution of A/I means that

$$H(K(I)) \approx A/I \quad \text{in ghost degree zero}$$

and $\qquad = 0 \quad$ otherwise.

Having introduced $K(I)$, we can now substitute it for A/I and consider

$$\mathrm{Alt}_A(I,K(I))$$

for which the homology with respect to d_K is $\mathrm{Alt}_A(I,A/I)$. If we want to pull the Cartan-Chevalley-Eilenberg differential over, however, the obvious representation of I on $A \subset K(I)$ will not do because it does not commute with d_K :

$$\{\gamma_\beta, f \otimes \rho_\alpha\} = \{\gamma_\beta, f\} \otimes \rho_\alpha$$

$$d_K\{\gamma_\beta, f \otimes \rho_\alpha\} = \{\gamma_\beta, f\}\gamma_\alpha$$

but $$\{\gamma_\beta, d_K(f \otimes \rho_\alpha)\} = \{\gamma_\beta, f\gamma_\alpha\} = \{\gamma_\beta, f\}\gamma_\alpha + f\{\gamma_\alpha, \gamma_\beta\} .$$

Avoiding this difficulty, FBV go further; they build a complex of the form

$$\Lambda\eta^\alpha \otimes A \otimes \Lambda\rho_\alpha$$

where the η^α are called anti-ghosts and have ghost degree -1. If there are infinitely many ghosts, it would be better for the above notation to be an abuse, really meaning

$$\mathrm{Alt}_A(\Phi, A \otimes \Lambda\rho_\alpha) ,$$

where Φ is the real span of the γ_α . Not only have FBV produced a much smaller complex than $\mathrm{Alt}(I,K(I))$, they have overcome the following:

Problem: Φ is NOT closed under $\{,\}$, (in physics, it is called an open algebra), unless

$$\{\gamma_\alpha, \gamma_\beta\} = C^\gamma_{\alpha\beta}\, \gamma_\gamma$$

with $C^\gamma_{\alpha\beta}$ being constants. Similarly, $\theta = \mathrm{ad}(\ ) : \Phi \to \mathrm{Aut}(A/I)$ is not closed. This results in $(d_k + d_\Phi + d_\theta)^2 \neq 0$. What to do?

FBV work in the alternate notation

$$d_K \longleftrightarrow \gamma_\alpha \eta^\alpha$$

$$d_\Phi + d_\theta \longleftrightarrow -\frac{1}{2} C^\gamma_{\alpha\beta}\, \eta^\alpha \eta^\beta \rho_\gamma + \{\gamma_\alpha, \ \}\eta^\alpha .$$

They extend $\{\ ,\ \}$ to $\Lambda\eta^\alpha \otimes A \otimes \Lambda\rho_\alpha$ in the standard way for Lie algebras: $\{\ ,\ \}$ on A and Λ-product otherwise, and then prove:

Theorem. There exist polynomials

$$U^{\alpha_1\ldots\alpha_i} \text{ in } \Lambda\eta^{\alpha} \otimes A \text{ such that}$$

$$U^{\alpha_1\ldots\alpha_i}\rho_{\alpha_1}\ldots\rho_{\alpha_i} \text{ has ghost degree 1 and}$$

$$\Omega = \varphi_{\alpha}\eta^{\alpha} + \{\varphi_{\alpha},\ \}\eta^{\alpha} + \sum_{I} U^{\alpha_I}\rho_{\alpha_I}\ , \ I = \alpha_1 < \alpha_2 \ldots < \alpha_i$$

satisfies

$$\{\Omega,\Omega\} = 0 \ .$$

In differential notation, this is equivalent to:

Theorem: There exist derivations δ_i of ghost degree 1 which increase the number of ρ's by i such that $D^2 = 0$ for $D = d_K + \delta_{\Phi} + \delta_{\theta} + \delta_1 + \delta_2 + \ldots$.

Fradkin et al find these terms of higher order in succession by setting up and solving a system of ODE's. Henneaux makes use of the resolution property of $K(I)$: each U^{α_I} exists because the obstruction to its existence can be computed to be a cycle which in $K(I)$ is automatically a boundary. The computations are quite complex and, at each stage, choices must be made. Looking at McMullan's exposition inspired me to apply the techniques of homological perturbation theory, and thus to see that only one choice is necessary (a contracting homotopy for $K(I)$) and the existence of terms of higher order is guaranteed by the general results of homological perturbation theory.

The word perturbation is due to one of the originators, V. K. A. M. Gugenheim [10], inspired by the term as used in physics, but not in his wildest dreams did he believe his theory would apply to physics!

Homological perturbation theory has developed gradually in a series of papers. Essential points in its development are in the papers of Gugenheim [10] and in his papers with May [11] and Stasheff [12]. Hopefully, Gugenheim, Lambe and Stasheff will soon have a summary and exposition of the fully developed theory. Here at the Institute, the local practitioner is Huebschmann [14].

In essence, the theory involves strong deformation retraction (SDR) data:

$$(C \underset{f}{\overset{\nabla}{\rightleftarrows}} E, h)$$

consisting of differential graded objects and differential graded morphisms with $f\nabla = id_C$ and h being a homotopy between ∇f and id_E. If either C (in our case) or E has another differential δ with $(d_C+\delta)^2 = 0$, then $(d_E+\nabla\delta f)$ need not have square zero but terms of higher order δ_i are guaranteed so that $(d_E + \nabla\delta f + \delta_1 + \ldots)$ does have square zero and with the extended differentials, C and D still have isomorphic homology.

To apply this machinery to produce the FBV complex, we have:

Theorem (Mike Schlessinger): $Alt_A(I, A/I) \approx Alt_{A/I}(I/I^2, A/I)$ and if I is a regular ideal,

$$Alt_{A/I}(I/I^2, A/I) \approx Alt_{\mathbb{R}}(\Phi, A/I) .$$

The first isomorphism is a standard argument with quotients and the second is effectively a characterization of Φ as a regular basis for I.

Steve Halperin has (since the original talk) called my attention to the work of G. S. Rinehart on "Differential forms on general commutative algebras" [18]. The Poisson bracket induces a representation of the Lie algebra I/I^2 on A/I by derivations and I/I^2 is an A/I module, so in Rinehart's terminology, I/I^2 is an $(\mathbb{R}, A/I)$-Lie algebra and the homology we study is denoted by $H_{A/I}(I/I^2; A/I)$.

Applying the above isomorphisms, we have an induced differential on $Alt(\Phi, A/I)$. If we replace A/I by the Koszul complex, we have exactly the SDR-data which enables us to conclude that the Cartan-Chevalley-Eilenberg differential on $Alt_A(I, A/I)$ gives rise to the FBV differential. An extension of the Basic Perturbation Lemma [15] is needed since we require the initial terms to be $\delta_\Phi + \delta_\theta$.

Returning to the physical interpretation, do the arbitrary choices imply the theory is ambiguous? No, if we regard $Alt(\Phi, A \otimes \Lambda\Phi)$ as an extended phase space, then the choices correspond precisely to a "canonical" change of coordinates: $(p,q) \longrightarrow (p',q')$.

What, then, do the terms of higher order signify? This is easiest to explain in the special case in which the total differential can be decomposed as $d_K + d_\Phi + d_\theta$ where $d_K + d_\Phi$ is a perturbation of $d_K + d_\Phi$ on $\mathrm{Alt}(\Phi,R)$. In this case, δ_1 tells us that $\{\ ,\ \}$ satisfies Jacobi modulo the image of d_K , i.e., "up to homotopy". The higher δ_i gives us analogous identities up to homotopy for four and more variables.

The corresponding structure at the group (rather than Lie algebra) level is easier to picture: Denoting a homotopy for associativity by $(xy)z$ — $x(yz)$, we can apply the homotopy to four variables in five ways which form a pentagon

$$\begin{array}{ccccc} & w((xy)z) & \text{---} & (w(xy))z & \\ \diagup & & & & \diagdown \\ w(x(yz)) & & & & ((wx)y)z \\ \diagdown & & & & \diagup \\ & & (wx)(yz) & & \end{array}$$

The existence of δ_2 corresponds to the contractability of this loop.

Similarly, the terms in D_Φ show that θ is a representation up to homotopy in a strong multivariable sense, beginning with

$$\theta([X,Y]) - \theta(X)\theta(Y) + \theta(Y)\theta(X) = d_K\theta_2(X,Y) \ .$$

Both of these homotopy structures imply the existence of multivariable operations in homology, i.e., Massey brackets [8].

Of course, if $C^\gamma_{\alpha\beta}$ were constants to begin with, none of this would have been necessary. This is very much NOT the case for gravity, though FBV claim that in (their model of) gravity, only δ_1 is non-zero!

In contrast, the work of Berends, Burgers and van Dam [4,7] on massless spin 3 particles indicates that there no finite perturbation will have square zero.

So far, I've talked entirely in terms of constraints. The Hamiltonian formalism refers to a dynamics given by the differential equation

$$\frac{d\Psi}{dt} = \{H,\Psi\} \ .$$

A physical problem begins with $H \in A = C^\infty(W)$ such that $\{H,\Psi_\alpha\} \in I$ for all constraints Ψ_α. To treat $\mathrm{Alt}(\Phi, A \otimes \Lambda\Phi)$ as an extended phase space, we need to extend H to $\bar{H} \in \mathrm{Alt}(\Phi, A \otimes \Lambda\Phi)$ of total ghost degree zero with $D\bar{H} = 0$. Again, this is precisely what is guaranteed by homological perturbation theory.

What about quantization? Everything seems to go through smoothly. Observables are identified with elements of the homology in total ghost degree zero of $\mathrm{Alt}(\Phi, A \otimes \Lambda\Phi)$, while states correspond to the corresponding homology of the L_2-dual. The ghosts ρ_α and antighosts η^α are then implemented as operators on (representatives of) states:

$$\hat{\eta}^\alpha = \text{multiplication by } \eta^\alpha$$

$$\bar{\rho}_\alpha = -i\partial/\partial\eta^\alpha .$$

Physical states are represented by $\bar{D}$ cycles: $\bar{D}|\Psi\rangle = 0$. (The operator $\bar{D}$ is obtained from $D = \Psi_\alpha\eta^\alpha + \{\Psi_\alpha, \ \}\eta^\alpha + U^{\alpha}{}_{I}\rho_I$ by $\eta \to \hat{\eta}, \rho \to \bar{\rho}$.) Two representatives Ψ_1, Ψ_2 are physically equivalent if $\Psi_1 - \Psi_2 = \hat{D}x$.

As elsewhere in contemporary physics, it is time to treat seriously the category of differential graded Hilbert spaces.

Bibliography

[1] I.A. Batalin and G.S. Vilkoviasky, *Existence theorem for gauge algebra*, J. Math. Phys. 26 (1985) 172-184.

[2] ——————, *Relativitistic S-matrix of dynamical systems with bosons and fermion constraints*, Physics Letters 69B (1977) 309-312.

[3] C. Becchi, A. Rouet and R. Stora, *Renormalization of gauge theories;* Ann. Phys. 98 (1976), 287.

[4] F.A. Berends, G.J.H. Burgers and H. van Dam, *On the theoretical problems in constructing interactions involving higher spin massless particles*, preprint IFP234-UNC, 1984

[5] L. Bonora and P. Cotta-Ramusino, *Some remarks on BRS transformations, Anomalies and the Cohomology of the Lie algebra of the Group of Gauge Transformations*, Comm.Math.Phys. 87 (1983), 589.

[6] A.D. Browning and D. McMullan, *The Batalin, Fradkin, Vilkovisky formalism for higher order theories*, preprint.

[7] G.J.H. Burgers, *On the construction of field theories for higher spin massless particles*, doctoral dissertation, Rijksuniversiteit te Leiden, 1985.

[8] A. Douady, *Obstruction primaire a la deformation, Seminaire Henri Cartan 1960/61, n° 4.*

[9] E.S. Fradkin and G.S. Vilkovisky, *Quantization of relativistic systems with constraints*, Physics Letters 55B (1975), 224-226.

[10] V.K.A.M. Gugenheim, *On a perturbation theory for the homology of a loop space*, J. Pure and Appl. Alg. 25 (1982), 197-205.

[11] V.K.A.M. Gugenheim and J.P May, *On the theory and application of torsion products*, Mem. Amer. Math. Soc. 142 (1974).

[12] V.K.A.M. Gugenheim and J. Stasheff, *On Perturbations and A-structures*, to appear in Proc. Soc. Math. Belg. volume in honor of Guy Hirsch.

[13] M. Henneaux, *Hamiltonian form of the path integral for theories with a gauge freedom* , Phy. Rep. 126 (1985) 1-66.

[14] J. Huebschmann, Perturbation Theory and Small Models for the Chains of Certain Induced Fibre Spaces, Habilitationsschrift, 1983, Heidelberg.

[15] L. Lambe and J. Stasheff, *Perturbation theory for iterated principal bundles*, preprint.

[16] D. McMullan, *Yang-Mills theory and the Batalin Fradkin Vilkovisky formalism*, preprint.

[17] P. Perigrinus, in A Source Book in Medieval Science, E. Grant, ed., Harvard Univ. Press, Cambridge (1974), p. 368.

[18] G. S. Rinehart, *Differential forms on general commutative algebras*, Trans. Amer. Math. Soc. 108 (1963), 195-222.

THE INDEX OF THE DIRAC OPERATOR IN LOOP SPACE

Edward Witten*

Joseph Henry Laboratories
Princeton University
Princeton, New Jersey 08544

1. Introduction

Let M be a spin manifold and $\mathcal{L}M$ the free loop space of maps $S^1 \to M$. Although analysis on infinite dimensional spaces such as $\mathcal{L}M$ is a new subject in the mathematical literature, it has been in the last sixty years much studied by physicists in the framework of quantum field theory. In particular, certain quantum field theories can be interpreted as infinite dimensional analogues of structures which have an established mathematical significance in the finite dimensional case. For instance, I showed in [1] that the supercharge Q of the supersymmetric nonlinear sigma model in $1+1$ dimensions (with target space M) can be interpreted as a Dirac-like operator on $\mathcal{L}M$.

These lecture notes will be devoted to discussing formally the index of the Dirac operator on $\mathcal{L}M$. This subject is closely related to the recent work of Schellekens and Warner on anomalies [2], and it has interesting applications, as we will see, to a certain topological problem which has been discussed at this conference by Landweber and Ochanine. In essence we will see that the topological conjecture in question would follow from certain simple (conjectured) properties of the supersymmetic nonlinear sigma model. Since a cutoff version of the nonlinear sigma model would be adequate, there is a reasonable hope that the requisite properties of the sigma model can be proved within a few years. This application of the sigma model to topology was briefly discussed in [3], along with some speculations about applications to physics. Our considerations are also closely related to the use of fixed point theorems on the loop space of a group to obtain the Weyl-Kac character formula for affine Lie algebras; see [4] for an exposition of this subject.

2. The Dirac Index on $\mathcal{L}M$

Let M be an even dimensional spin manifold, and

$$D : \Gamma(S_+) \to \Gamma(S_-) \tag{1}$$

the Dirac operator (S_+ and S_- are the positive and negative spin bundles and $\Gamma(\cdots)$ denotes the

* Research supported in part by NSF Grants PHY80-19754 and 86-16129

smooth sections of a vector bundle). The index of the Dirac operator is defined as

$$\operatorname{index} D \;=\; \dim H_+ \;-\; \dim H_- \tag{2}$$

where H_+ and H_- are the kernel and cokernel of D. If G is a compact group that acts on M, and we pick a G-invariant metric on M, then H_+ and H_- become G modules. For $g \epsilon G$ we then define

$$F(g) \;=\; Tr_{H_+}\, g \;-\; Tr_{H_-}\, g. \tag{3}$$

The function $F(g)$ obviously depends only on the conjugacy class of $g \epsilon G$ and is known as the character-valued Dirac index. Evidently, $F(1)$ is the Dirac index in the sense of equation (2).

The index theorem leads to a formula that determines $F(g)$ in terms of the fixed points of G [5]. Given $g \epsilon G$, if the fixed point set M^g of g is a union of connected components M^g_α, the fixed point formula is of the general nature

$$F(g) \;=\; \sum_\alpha F_\alpha(g) \tag{4}$$

where $F_\alpha(g)$ depends on local data at M^g_α.

Let us specialize to the case $G = S^1$, which will be of particular interest for us. The general element of G is then

$$g \;=\; e^{\theta P} \tag{5}$$

with P a generator of S^1, and θ an angular variable, $0 \leq \theta \leq 2\pi$. The character-valued index becomes a function $F(\theta)$. Let us describe explicitly the fixed point formula for $F(\theta)$. Consider first the contribution of an isolated fixed point x.

In the tangent space T_x to M at x, one can find an orthonormal basis $e_1, \cdots, e_{2k}$ $(2k = \dim M)$ in which P takes the form

$$P \;=\; \begin{pmatrix} & n_1 & & & & & \\ -n_1 & & & & & & \\ & & & n_2 & & & \\ & & -n_2 & & & & \\ & & & & \ddots & & \\ & & & & & & n_k \\ & & & & & -n_k & \end{pmatrix} \tag{6}$$

with integers n_i, which are non-zero as x is an isolated fixed point.

The e_i can be chosen so the n_i are all positive. Fixing an orientation ϵ of M, which induces an orientation ϵ_x of each T_x, let λ_x be +1 or -1 according to whether $e_1 \wedge e_2 \cdots \wedge e_{2k}$ is a positive or negative multiple of ϵ_x. The contribution of x to $F(\theta)$ is then

$$F_x(\theta) = \lambda_x \cdot \prod_{i=1}^{k} \frac{e^{in_i\theta/2}}{1 - e^{in_i\theta}} \tag{7}$$

and if all fixed points are isolated, then $F(\theta) = \Sigma_x F_x(\theta)$.

The generalization of (7) to the case of fixed points which may not be isolated is as follows. Let M_α be an arbitrary connected component of the fixed point set of S^1. The normal bundle N to M_α in M has a decomposition

$$N = \bigoplus_{\ell \neq 0} N_\ell \tag{8}$$

with P acting on N_ℓ as multiplication by $i\ell$ (the ℓ are integers).

For any vector bundle V and complex number t, we will let

$$\frac{1}{1 - tV} = 1 \oplus tV \oplus t^2 S^2 V \oplus \cdots \oplus t^k S^k V + \cdots \tag{9}$$

where $S^k V$ denotes the kth symmetric tensor power. We will also write $\frac{1}{1-tV}$ as $S_t V$.

Let $N_+ = \bigoplus_{\ell > 0} N_\ell$, and let $\det N_+$ $(= \bigotimes_{\ell > 0} \det N_\ell)$ denote the highest exterior power. Let n_ℓ be the dimension of N_ℓ. The generalization of (7) is then

$$F_\alpha(\theta) = \lambda_\alpha \left\langle \hat{A}(M_\alpha) \operatorname{ch} \left\{ \sqrt{\det N_+} \cdot \prod_{\ell > 0} e^{i\theta \ell n_\ell / 2} \cdot \bigotimes_{\ell > 0} \frac{1}{1 - e^{i\ell\theta} N_\ell} \right\}, M_\alpha \right\rangle \tag{10}$$

(The combination $\sqrt{\det N_+} \cdot \prod_{\ell > 0} e^{i\theta \ell n_\ell / 2}$ should be viewed as the equivariant generalization of $\sqrt{\det N_+}$.) Here $\hat{A}(M_\alpha)$ is the total $\hat{A}$ class of M_α, and the sign λ_α is determined as before. The character-valued index of the Dirac operator is then $F(\theta) = \Sigma_\alpha F_\alpha(\theta)$. If $H_1(M_\alpha, Z_2) \neq 0$, then $\det N_+$ has various square roots and it is necessary to specify a consistent choice corresponding to a choice of spin structure on M.

All of this has the following generalization. Let V be a vector bundle on M to which the S^1 action on M has been lifted, and suppose $w_2(T) = w_2(V)$ (T being the tangent bundle of M). Then we can consider the character-valued index of a twisted Dirac operator

$$D_V : \Gamma(S_+ \otimes V) \longrightarrow \Gamma(S_- \otimes V) \tag{11}$$

The fixed point formula is now modified as follows. The restriction V_α of V to M_α will decompose as

$$V_\alpha = \bigoplus_\ell V_{\ell,\alpha} \tag{12}$$

where P acts on V_ℓ as multiplication by $i\ell$. (10) is then replaced by

$$\begin{aligned} F_\alpha^V(\theta) = \lambda_\alpha \Big\langle \hat{A}(M_\alpha)\mathrm{ch}\Big\{\sqrt{\det N_+}\cdot \prod_{\ell>0} e^{i\theta\ell n_\ell/2}\cdot \bigotimes_{\ell>0}\frac{1}{1-e^{i\ell\theta}N_\ell}\Big\} \\ \bigoplus_m e^{im\theta} V_{m,\alpha}\,,\ M_\alpha\Big\rangle \end{aligned} \tag{13}$$

Our goal is now to work out the formal analogues of (10) and (13) for the case of Dirac-like operators on the free loop space $\mathcal{L}M$ of maps $S^1 \to M$.

Regardless of the choice of M, $\mathcal{L}M$ always admits a natural S^1 action. Indeed, if σ is a standard angular parameter on S^1 (so $0 \le \sigma \le 2\pi$) then the translation $\sigma \to \sigma + c$, c a constant, induces an S^1 action on $\mathcal{L}M$. It is the character-valued index for this group action on $\mathcal{L}M$ that we will now study.

If X^i are coordinates for M, a map $S^1 \to M$ may be described explicitly in terms of functions $X^i(\sigma)$. To have a fixed point for the natural S^1 action on $\mathcal{L}M$ means that $X^i(\sigma)$ must be invariant under the translation $\sigma \to \sigma+c$; *i.e.* it must be a constant map $X^i(\sigma) = X_0^i$, with X_0^i the coordinates for a point in M. Thus, a fixed point of the S^1 action is simply a constant map $S^1 \to M$, and the fixed point set is just a copy of M, embedded in $\mathcal{L}M$. Our basic idea in these notes is to study M via this embedding in $\mathcal{L}M$, and we henceforth identify M with its image in $\mathcal{L}M$.

To describe the normal bundle to M in $\mathcal{L}M$, we note that an almost constant map $S^1 \to M$ is of the form

$$X^i(\sigma) = X_0^i + \sum_{n\neq 0} e^{in\sigma}\epsilon_n^i \tag{14}$$

Here, for each non-zero integer n, ϵ_n^i is a vector tangent to M at the point with coordinates X_0^i. Thus, the decomposition of the normal bundle of M in $\mathcal{L}M$ (analogous to (8)) is

$$N = \bigoplus_{\ell\neq 0} T_\ell \tag{15}$$

where the T_ℓ are all isomorphic to the tangent bundle T of M.

We may now readily determine the analogue of equation (10). As M is a spin manifold, and in particular orientable, $\det T$ is a trivial line bundle, and $\det N_+$, an infinite tensor product of trivial line bundles, can be dropped. ($\sqrt{\det N_+}$ is an arbitrary flat line bundle of order two, corresponding to a choice of spin structure on M.) The n_ℓ are all equal to $d = 2k = \dim M$. The analogue of $\Pi_{\ell>0}\, e^{i\theta\ell n_\ell/2}$ is thus

$$\left(\prod_{n=1}^{\infty} q^n\right)^{d/2} \tag{16}$$

where we have let $q = e^{i\theta}$. We interpret $\prod_{n=1}^{\infty} q^n = q^{\sum_{n=1}^{\infty} n}$ as $q^{\zeta(-1)} = q^{-1/12}$. Thus (16) becomes $q^{-d/24}$. The remaining factor in (10) is just $\bigotimes_{\ell=1}^{\infty} \frac{1}{1-q^\ell T} = \bigotimes_{\ell=1}^{\infty} S_{q^\ell}T$. Thus, the formal expression for the character-valued Dirac index on $\mathcal{L}M$ is

$$F(q) \;=\; q^{-d/24} \left\langle \hat{A}(M)\mathrm{ch} \bigotimes_{\ell=1}^{\infty} S_{q^\ell}T \,,\, M \right\rangle \tag{17}$$

The reason that the variable in (17) has been called q is as follows. It can be shown that (17) is of the form

$$\frac{\Phi(q)}{\eta(q)^d} \tag{18}$$

with $\eta(q) = q^{1/24}\prod_{\ell=1}^{\infty}(1-q^\ell)$ the Dedekind eta function and $\Phi(q)$ a modular form of weight $d/2$ for $SL(2,Z)$, provided $p_1(M) = 0$. The modular form $\Phi(q)$ is the so-called level one elliptic genus of M. The restriction on p_1 has to do with anomalies, and will be explained heuristically in the next section. The natural transformation law of (17) under $SL(2,Z)$, which may come as a surprise in the present exposition, has a simple explanation in terms of facts that are well known to physicists. Indeed, the Feynman path integral representation of $F(q)$ involves integration over maps $\Sigma \to M$, where Σ is an elliptic curve over C with complex structure determined by q; this representation leads naturally to an understanding of the behavior of (17) under modular transformations of q. I will not enter into this here (but see [6] for a qualitative discussion of an analogous question, namely the role of $SL(2,Z)$ in the theory of affine Lie algebras and in "monstrous moonshine"). A computational proof of the modular properties of (17) under the restriction $p_1(M) = 0$ can be given without reference to Feynman path integrals; see [2] and the article by Zagier in this volume.

We will now generalize (17) to a formal expression for the character-valued index of various twisted Dirac-like operators on $\mathcal{L}M$. The first case that we will consider is the signature operator on $\mathcal{L}M$. The signature operator is simply

$$D_S \,:\, \Gamma(S_+ \otimes S) \to \Gamma(S_- \otimes S) \tag{19}$$

where $S = S_+ \oplus S_-$ is the spin bundle. Thus we must construct the spin bundle S of $\mathcal{L}M$. Actually, since we are only dealing with fixed point formulas, we only need the restriction of S to $M \subset \mathcal{L}M$.

Let $\mathcal{T}$ be the tangent bundle of $\mathcal{L}M$, restricted to M. Then

$$\mathcal{T} = T \oplus N \tag{20}$$

with T the tangent bundle of M and N its normal bundle in $\mathcal{L}M$. For any real vector bundle A, let $\Delta(A)$ be the associated spinor bundle. Then $\Delta(A \oplus B) = \Delta(A) \otimes \Delta(B)$. Thus

$$S = \Delta(\mathcal{T}) = \Delta(T) \otimes \Delta(N) \tag{21}$$

We decompose N as in equation (15). It is convenient to combine T_ℓ and $T_{-\ell}$ together. So $N = \oplus_{\ell>0} K_\ell$ with $K_\ell = T_{-\ell} \oplus T_\ell$, and

$$\Delta(N) = \bigotimes_{\ell=1}^{\infty} \Delta(K_\ell) \tag{22}$$

In general, if $\bar{A}$ denotes the dual bundle of A, then

$$\Delta(A \oplus \bar{A}) = \sqrt{\det \bar{A}} \cdot \Lambda(A)$$

where $\Lambda(A) = 1 \oplus A \oplus \Lambda^2 A \oplus \cdots$ is the direct sum of the exterior powers. (For any vector space A, $\Lambda^k A$ will denote its kth exterior power.) With $K_\ell = T_{-\ell} \oplus T_\ell$ (and $T_{\pm\ell}$ both isomorphic to T, so that $\sqrt{\det T_{-\ell}}$ is trivial), this shows that $\Delta(K_\ell) = \Lambda(T)$. We need, however, a more refined formula that keeps track of the S^1 action. With $K_\ell = q^{-\ell} T_{-\ell} \oplus q^\ell T_\ell$ this formula is

$$\Delta(K_\ell) = q^{-d\ell/2} \Lambda_{q^\ell} T \tag{23}$$

The spinor bundle of $\mathcal{L}M$, restricted to M, is thus

$$\Delta(\mathcal{T})|_M = q^{\frac{d}{24}} \Delta(T) \bigotimes_{\ell=1}^{\infty} \Lambda_{q^\ell} T \tag{24}$$

Here $\Lambda_t T$ represents $1 \oplus tT \oplus t^2 \Lambda^2 T \oplus \cdots$. More precisely, this is a description of what we *mean* by the spinor bundle of $\mathcal{L}M$; a priori, there are various conceivable choices. Putting (24) together with (13) and (17), we get a formal expression for the equivariant signature of $\mathcal{L}M$:

$$G(q) = \left\langle \hat{A}(M) \mathrm{ch} \bigotimes_{\ell=1}^{\infty} S_{q^\ell} T \bigotimes \Delta(T) \bigotimes_{m=1}^{\infty} \Lambda_{q^m} T \, , \, M \right\rangle \tag{25}$$

The Feynman path integral representation for (25) shows that it is of the form

$$G(q) = \Phi(q) \left(\frac{\eta(q^2)}{\eta^2(q)} \right)^d \tag{26}$$

where $\Phi(q)$ is a modular form for a level two subgroup Γ of $SL(2, Z)$ which has index three in $SL(2, Z)$. (There is no requirement here on $p_1(M)$.) The "level two" property arises because the

Feynman path integral representation of (25) involves integrating over maps $\Sigma \to M$ where now the elliptic curve Σ is endowed with a point of order two. As $[SL(2,Z) : \Gamma] = 3$, (25) can be transformed into two essentially different formulas by $SL(2,Z)$ transformations not in Γ. One of these is

$$H(q) = q^{-d/16} \left\langle \hat{A}(M) \bigotimes_{\ell=1}^{\infty} S_{q^\ell} T \bigotimes_{m=1/2,3/2,5/2,\cdots} \Lambda_{q^m} T \,,\, M \right\rangle \tag{27}$$

The other differs by $q^{1/2} \to -q^{1/2}$. (27) can be interpreted as the index of a sort of twisted version of the signature operator on $\mathcal{L}M$. The operator in question, while well known to physicists (it is the supercharge with right-moving Ramond and left-moving Neveu-Schwarz boundary conditions), does not have a finite dimensional analogue. Its existence is a characteristic feature of analysis on the infinite dimensional manifold $\mathcal{L}M$. I will now describe it briefly.

A point γ in $\mathcal{L}M$ is a loop

$$\gamma : S^1 \to M \tag{28}$$

Pulling back the tangent bundle T of M via γ gives a vector bundle γ^*T over S^1. Let T_γ be the space of sections of γ^*T. The family $\{T_\gamma | \gamma \epsilon \mathcal{L}M\}$ form the fibers of an infinite dimensional vector bundle $\mathcal{T}$ over $\mathcal{L}M$; it is none other than the tangent bundle of $\mathcal{L}M$, which we discussed earlier.

Now, let ϵ be the Hopf bundle of S^1 —the unique non-trivial real line bundle over S^1. Let $\hat{T}_\gamma$ be the space of sections of $\epsilon \otimes \gamma^*T$. The family $\{\hat{T}_\gamma | \gamma \epsilon \mathcal{L}M\}$ are fibers of a vector bundle over $\mathcal{L}M$ which we may call $\hat{\mathcal{T}}$.

The restriction of $\hat{\mathcal{T}}$ to $M \subset \mathcal{L}M$ is particularly simple; it is

$$\hat{\mathcal{T}}|_M \simeq \bigoplus_{m \epsilon Z + \frac{1}{2}} q^m T_m \tag{29}$$

with T_m isomorphic to T for all $m \epsilon Z + \frac{1}{2}$. The factors q^m in (29) have been introduced to keep track of the S^1 action. From (29) one finds the analogue of (24):

$$\Delta(\hat{\mathcal{T}})|_M = q^{-\frac{d}{48}} \bigotimes_{m=1/2,3/2,5/2,\cdots} \Lambda_{q^m} T \tag{30}$$

From here one obtains (27) as the index of the Dirac operator on $\mathcal{L}M$ twisted by $\Delta(\hat{\mathcal{T}})$. The Feynman path integral gives a conceptual explanation, well known among string theorists, of the fact that the Dirac operator twisted by $\Delta(\mathcal{T})$ is related by $SL(2,Z)$ to that twisted by $\Delta(\hat{\mathcal{T}})$. The fact that (25) and (27) are related by $SL(2,Z)$ can also be verified computationally.

Now, I wish to consider some generalizations that depend on the choice of a vector bundle V on M. For reasons that will be discussed qualitatively in the next section, a good theory will emerge only if $w_2(V) = w_2(T)$ and $p_1(V) = p_1(T)$.*

Considering again a loop $\gamma : S^1 \to M$, we pull back V to $\gamma^* V$, and let V_γ denote the space of sections of $\gamma^* V$. Let $\mathcal{V}$ be the vector bundle over $\mathcal{L}M$ whose fiber at $\gamma \epsilon \mathcal{L}M$ is V_γ. The index of the Dirac operator on $\mathcal{L}M$ twisted by $\Delta(\mathcal{V})$ is

$$F_V(q) = q^{-d/24} q^{n/24} \left\langle \hat{A}(M) \mathrm{ch} \bigotimes_{\ell=1}^{\infty} S_{q^\ell} T \bigotimes \Delta(V) \bigotimes_{m=1}^{\infty} \Lambda_{q^m} V, M \right\rangle \tag{31}$$

as one can see by repeating the arguments that led to (25). Here n is the dimension of V. Equation (31) is of the form

$$\Phi(q) \cdot \frac{\eta(q^2)^d}{\eta(q)^{d+n}} \tag{32}$$

with $\Phi(q)$ a modular form of weight $d/2$ for a level two subgroup of $SL(2,Z)$, if $p_1(V) = p_1(T)$.

Finally, if V is even dimensional, then we can make a decomposition

$$\Delta(\mathcal{V}) = \Delta_+(\mathcal{V}) \oplus \Delta_-(\mathcal{V}) \tag{33}$$

analogous to the decomposition of the spin bundle in finite dimensions. Then we consider the Dirac index twisted by $\Delta_+(\mathcal{V}) \ominus \Delta_-(\mathcal{V})$ instead of $\Delta_+(\mathcal{V}) \oplus \Delta_-(\mathcal{V})$ to get

$$J_V(q) = q^{-\frac{(d-n)}{24}} \left\langle \hat{A}(M) \mathrm{ch} \bigotimes_{\ell=1}^{\infty} S_{q^\ell} T \bigotimes (\Delta_+(V) \ominus \Delta_-(V)) \bigotimes_{m=1}^{\infty} \Lambda_{-q^m} V, M \right\rangle \tag{34}$$

If $n = \dim V$ exceeds $d = \dim M$, this is zero. For $n = d$, $J_V(q)$ is independent of q and equals the Euler characteristic of V. For $n < d$, (34) is of the form

$$\frac{\Phi(q)}{\eta(q)^{d-n}} \tag{35}$$

with $\Phi(q)$ a modular form of weight $(d-n)/2$ for $SL(2,Z)$. In contrast to our previous examples, (34) is an unstable characteristic class which might be regarded as the analogue in elliptic cohomology of the Euler characteristic of a vector bundle. Despite being unstable, (34) is the subject of an interesting mathematical theory; the theorem that we will discuss in section 4 concerning the vanishing of equivariant characteristic numbers applies to (34) as well as (31).

---

* For a real vector bundle (or virtual bundle such as $T \ominus V$) with $w_2 = 0$, there is a natural way to define $p_1/2$ (*i.e.*, there is a characteristic class λ with $2\lambda = p_1$), and the equality $p_1(V) = p_1(T)$ must be taken to include this twofold refinement.

3. Anomalies

In this section we will indicate a few aspects of the relation between anomalies and the Dirac operator in loop space. The whole subject of anomalies is quite vast, and we will content ourselves with drawing attention to a few relevant points. Our main purpose is to indicate why the (untwisted) Dirac operator on $\mathcal{L}M$ only makes sense if $p_1(M) = 0$. This was originally discovered at the rational level in [7]. In [1], global anomalies were used to obtain the restriction on p_1 (or actually $p_1/2$) as an *integral* cohomology class. In my brief remarks here, I will follow a viewpoint proposed by Killingback [8]; I refer to his paper for more detail.

Let M be a manifold of finite dimension n. The structure group of the tangent bundle T is $SO(n)$. The spinor bundle $\Delta(T)$, if it exists, has structure group Spin (n), this being the simply connected double cover of $SO(n)$:

$$0 \to Z_2 \to \mathrm{Spin}(n) \to SO(n) \to 0 \tag{36}$$

As Z_2 is a discrete group, the group extension in (36) is trivial at the Lie algebra level, and shows up only globally. The obstruction to lifting from $SO(n)$ to $\mathrm{Spin}(n)$ involves a two dimensional cohomology class, $w_2(M)$, which must vanish if $\Delta(T)$ is to exist.

Now let us consider the analogous issues on $\mathcal{L}M$, the loop space of the n dimensional manifold M. The structure group of the tangent bundle $\mathcal{T}$ of $\mathcal{L}M$ is naturally $\mathcal{L}SO(n)$ — the loop group of $SO(n)$. The fundamental difference between the finite and infinite dimensional cases arises when one attempts to construct the spinor bundle $\Delta(\mathcal{T})$. Constructing $\Delta(\mathcal{T})$ will of course involve a central extension of $\mathcal{L}SO(n)$, as in (36). The essential novelty of loop space is that the central extension that arises is by the Lie group $U(1)$ rather than the discrete group Z_2:

$$0 \to U(1) \to \mathcal{G} \to \mathcal{L}SO(n) \to 0 \tag{37}$$

The Lie algebra corresponding to $\mathcal{G}$ is the so-called affine Lie algebra $\widehat{SO}(n)$; up to normalization of the central generator it is the unique non-trivial central extension of the Lie algebra of $\mathcal{L}SO(n)$. It is $\mathcal{G}$ which is the structure group of the spinor bundle $\Delta(\mathcal{T})$ of $\mathcal{L}M$, as this spinor bundle was described in the last section.

By way of explaining that last statement, I will only observe the following. We have already described in equation (24) the restriction of $\Delta(\mathcal{T})$ to $M \subset \mathcal{L}M$:

$$\Delta(\mathcal{T})|_M = q^{d/24}\,\Delta(T)\bigotimes_{\ell=1}^{\infty}\Delta_{q^\ell}T \tag{38}$$

From the construction of $\Delta(\mathcal{T})|_M$, one might expect that the Lie algebra of $\mathcal{L}SO(n)$ would act naturally on $\Delta(\mathcal{T}|_M)$. Instead, in trying to implement in the infinite dimensional context standard formulas for the spin representation, one finds that the Lie algebra that acts naturally on the module (38) is the central extension $\widehat{SO}(n)$ [9]. Indeed, (38) is a description of the fundamental spin representation of $\widehat{SO}(n)$; it represents in a way the most elementary construction of a highest weight module for any affine Lie algebra.

Anyway, the fact that the extension in (37) is an extension of $\mathcal{L}SO(n)$ by $U(1)$ means that the obstruction to the existence of $\Delta(\mathcal{T})$ is not a torsion element as in finite dimensions but a cohomology class with $U(1)$ coefficients. In fact, the projective spin bundle $P\Delta(\mathcal{T})$ (the bundle whose fiber at $\gamma \epsilon \mathcal{L}M$ is the complex projective space consisting of lines through the origin in the fiber at γ of $\Delta(\mathcal{T})$) always exists, regardless of the topology of M. Given $P\Delta(\mathcal{T})$, the obstruction to constructing $\Delta(\mathcal{T})$ is a three dimensional cohomology class on $\mathcal{L}M$. This in turn is related by transgression to a four dimensional cohomology class on M, namely $\frac{1}{2}p_1(M)$. Vanishing of $p_1(M)$ permits one to define the spinor bundle $\Delta(\mathcal{T})$ and thus the Dirac operator on $\mathcal{L}M$.* For elucidation of this I refer again to [8].

Now, let us briefly consider some of the twisted Dirac operators that figured in section 2. First, we consider the signature operator. On a finite dimensional manifold M, the signature operator σ can be defined if and only if M is orientable. Thus, to discuss the signature operator on $\mathcal{L}M$, we must know whether $\mathcal{L}M$ is orientable. I argued in [1] that $\mathcal{L}M$ should be considered orientable precisely if M is a spin manifold. This in any case is definitely the criterion that is relevant in quantum field theory. The quantum field theory that leads to the index (25) makes sense if and only if M is spin, and thus even though the formula (25) for the so-called Jacobi elliptic genus makes sense whenever M is orientable, any property of this formula that is proved using quantum field theory will require a spin condition.

In (31) we generalized the index of the signature operator to an index of a Dirac operator twisted by $\Delta(\mathcal{V})$, $\mathcal{V}$ being a bundle on $\mathcal{L}M$ obtained from an underlying vector bundle V on M. In this construction it is the tensor product of $\Delta(\mathcal{T})$ with the dual of $\Delta(\mathcal{V})$ that must exist, so the necessary requirement is $p_1(V) = p_1(T)$. The same is true for the "Euler characteristic" (34). Thus, our basic statement in the next section about equivariant constancy of (31) and (34) will require $p_1(V) = p_1(T)$.

Finally, with an eye toward the next section, we must discuss the equivariant generalization of the above, to a situation in which a group G is acting on M. We have then a fibration

$$\begin{array}{ccc} M & \longrightarrow & X \\ & & \downarrow \pi \\ & & BG \end{array}$$

over the classifying space BG. Let T_G be the tangent bundle to the fibers of this fibration.

It is possible to define a spin bundle $\Delta(T)$ and a Dirac equation on the manifold M if $w_2(T)$, the second Stiefel-Whitney class of the tangent bundle T, vanishes (one often denotes $w_2(T)$ as $w_2(M)$).

---

* It is conceivable but not likely that these can be defined if the following somewhat weaker condition is obeyed. If $X \subset M$ is a four dimensional cycle that can be fibered over a circle, then $< p_1(M), X >$ should vanish.

Given a group G that acts on a spin manifold M, it is not necessarily possible to choose on M a spin structure that admits an action of G, and thus G cannot necessarily act as a symmetry of the Dirac equation. The condition for being able to choose a G-equivariant spin bundle is that $w_2(T)$ must vanish in the equivariant sense. In fact, one needs $w_2(T_G) = 0$, with T_G the vector bundle described in the previous paragraph. More generally, if we are given a manifold M and a vector bundle V, we can define a Dirac equation for spinors twisted by V if and only if $w_2(T) = w_2(V)$. And in an equivariant situation, with a group G acting on M, T, and V, the Dirac equation for spinors twisted by V can be chosen to admit a G action if $w_2(T_G) = w_2(V_G)$. The definition of V_G is analogous to that of T_G : given the vector bundle V over M with G action, there is a natural vector bundle V_G over $X \to BG$ whose restriction to each fiber is isomorphic to V.

In the next section, we will study the Dirac operator on $\mathcal{L}M$ in a situation in which a group G is acting on M. Thus, we will need the generalization to $\mathcal{L}M$ of the above remarks. We already know that the Dirac operator on $\mathcal{L}M$ makes sense only if $p_1(T) = 0$ (or $p_1(T) = p_1(V)$, if one has twisted by a vector bundle $\Delta(\mathcal{V})$ derived from V as in the previous section). The obvious equivariant generalization of that restriction would be to say that the Dirac equation on $\mathcal{L}M$ should admit a G action only if $p_1(T_G) = 0$ (or more generally $p_1(T_G) = p_1(V_G)$). In physical terms, this restriction arises because if $p_1(T) = p_1(V)$, so that the Dirac equation in loop space makes sense, but $p_1(T_G) \neq p_1(V_G)$, then the G symmetry is violated by anomalies in instanton amplitudes, that is, the contributions to the Feynman path integral from homotopically non-trivial maps $X : \Sigma \to M$ (Σ is an elliptic curve) do not respect the G symmetry.†

Therefore, in formulating in the next section a theorem about vanishing of certain equivariant characteristic numbers, we will require a hypothesis that $p_1(T_G) = p_1(V_G)$. A natural and general way to obey this restriction is to pick $V = T$. This amounts to studying the signature operator on $\mathcal{L}M$, whose equivariant index was given in equation (25). However, our results will be valid for any V such that $p_1(T_G) = p_1(V_G)$.

4. Circle Actions and Characteristic Numbers

In this section, I will sketch the application of Dirac operators in loop space —or in other words of supersymmetric nonlinear sigma models —to a topological problem involving the vanishing of certain equivariant characteristic numbers.

Let M be an n-dimensional spin manifold that admits the action of a compact connected Lie group G. In what follows, it will be sufficient to consider the case $G = S^1$, as the generalization of our statements to arbitrary G follows by considering suitable S^1 subgroups of G.

---

† It is conceivable that the restriction $p_1(T_G) = p_1(V_G)$ can be relaxed slightly along the lines of the previous footnote.

Let K be the vector field that generates the S^1 action, and L_K the corresponding Lie derivative. Let $S_\pm$ be the spin bundles and T the tangent bundle of M. For any representation R of Spin(n), let T_R be the corresponding associated bundle derived from T; thus, T_R is simply T if R is the n dimensional real vector representation of Spin(n).

Picking an S^1 invariant metric on M, we consider then the Dirac operator

$$D_R : \Gamma(S_+ \otimes T_R) \longrightarrow \Gamma(S_- \otimes T_R) \tag{39}$$

where $\Gamma(V)$ denotes the smooth sections of a vector bundle V. The index of D_R is the dimension of its kernel minus that of its cokernel.

Since D_R and L_K commute, the operator D_R can be restricted to eigenspaces of L_K. For any vector bundle V with S^1 action, let $\Gamma_k(V)$ denote the space of sections ψ of V with eigenvalue k of L_K. The values of k that appear will be integers or half-integers if the S^1 action is "even" or "odd". Let $D_R^{(k)}$ be the restricted Dirac operator

$$D_R^{(k)} : \Gamma_k(S_+ \otimes T_R) \longrightarrow \Gamma_k(S_- \otimes T_R) \tag{40}$$

and let $c_{R,k}$ be its index. We will be discussing some vanishing theorems for the $c_{R,k}$. It is often convenient to combine the $c_{R,k}$ in a function

$$f_R(\theta) = \Sigma\, e^{ik\theta}\, c_{R,k} \tag{41}$$

If R is the trivial representation, then the $c_{R,k}$ are zero for all k, by a theorem of Atiyah and Hirzebuch [10]. It seems that the trivial representation is the only representation with this property. If, however, R is the spin representation, then D_R is essentially the signature operator, whose zero eigenvalues are harmonic differential forms, and it is a relatively elementary result that $c_{R,k} = 0$ for $k \neq 0$. One is naturally led to ask whether there are other representations R with that property. Some years ago, I conjectured [11] that the $c_{R,k}$ for $k \neq 0$ all vanish if R is the vector representation of Spin(n).‡ In trying to understand this conjecture, Landweber and Stong generalized it [12] to a certain infinite series of representations, which turned out, through the efforts of Ochanine, the Chudnovskys, and Zagier, to be naturally described in terms of certain elliptic modular functions [13-15]. It has emerged that the series of representations in question correspond precisely to the signature operator in $\mathcal{L}M$ (free loop space) described in section 2. I would now like to sketch how the vanishing of $c_{R,k}$ for $k \neq 0$, for the relevant R, follows formally from simple properties of the Dirac operator on $\mathcal{L}M$ or, if you will, from certain standard conjectures about quantum field theory.

‡ This result was by no means desirable at the time, since it frustrated certain efforts to explain parity violation in weak interactions.

First, let us state a precise conjecture. We define a sequence R_n of representations of Spin(n) by the formula

$$\sum_{n=0}^{\infty} q^n R_n = \Delta(T) \bigotimes_{\ell=1}^{\infty} S_{q^\ell}(T) \bigotimes_{m=1}^{\infty} \Lambda_{q^m}(T) \tag{42}$$

Here T and $\Delta(T)$ are symbols that represent, respectively, the vector and spinor representations of Spin(n); one is to formally expand the right-hand side in powers of q, the coefficient of q^n being R_n. These representations are the ones that were found in section 2 to arise in the study of the signature operator in loop space. The S^1 equivariant signature of loop space is according to equation (25) precisely

$$G(q) = \sum_{n=0}^{\infty} q^n \operatorname{index}(R_n) \tag{43}$$

where index (R_n) denotes the index of the Dirac operator acting on sections of $S_+ \otimes R_n$. We consider here the signature operator rather than one of the other operators on $\mathcal{L}M$ considered in section 2 because, as sketched in the last section, the signature operator is canonically free of all anomalies.

We wish to consider now a situation in which we are given a non-trivial S^1 action on M, generated by a vector field K. We refine the quantity index(R_n) to an equivariant index, as indicated in our introductory discussion. Thus, we let

$$c_{n,k} = (\operatorname{index} D_{R_n})_{L_K=k}\,. \tag{44}$$

The integer $c_{n,k}$ is thus simply the index of the nth Dirac operator D_{R_n} restricted to the kth eigenspace of L_K, $\Gamma_k(S_+ \otimes R_n)$. The conjecture that we wish to discuss asserts that $c_{n,k} = 0$ for $k \neq 0$.

It is useful to form the generating functional

$$G(q,\theta) = \Sigma c_{n,k}\, q^n\, e^{ik\theta}\,. \tag{45}$$

Here $G(q,\theta)$ can be regarded as the $S^1 \times S^1$ equivariant signature of $\mathcal{L}M$, the first S^1 being the universal circle action on $\mathcal{L}M$ and the second one being the circle action on $\mathcal{L}M$ induced from the S^1 action on M generated by K. At $\theta = 0$, $G(q,\theta)$ reduces to $G(q)$ as defined earlier.

In what follows, the series on the right-hand side of (45) can be viewed as a purely formal series, whose convergence is not relevant. We will be making statements about the integers $c_{n,k}$ that were concretely defined in equation (44).

The statement that $c_{n,k} = 0$ for $k \neq 0$ follows from two properties. One of these

$$c_{n,k} = 0 \quad \text{for} \quad n < 0 \tag{46}$$

is implicit in (42) and reflects at a more fundamental level the fact that the energy of the supersymmetric nonlinear sigma model is non-negative. The second property we will require is that

$$c_{n,k} = c_{n-k,k}. \tag{47}$$

This will follow from simple considerations involving the signature operator on a twisted loop space. From (47) one observes that $c_{n,k} = c_{n-tk,k}$ for any integer t, and choosing t so $|tk| > n$ it then follows from (46) that $c_{n,k} = 0$ for $k \neq 0$.

To prove (47) we introduce the "twisted loop space." Looking at S^1 as $R/2\pi Z$, R being the real line, we regard maps $S^1 \to M$ as maps $R \to M$ that are invariant under 2π translation. Thus, a map $\sigma \to X(\sigma)$ of $R \to M$ (σ and X denote points in R and M, respectively) induces a map $S^1 \to M$ precisely if

$$X(\sigma + 2\pi) = X(\sigma) \tag{48}$$

Now, for $g\epsilon S^1$, we define the twisted loop space $(\mathcal{L}M)^g$ to consist of maps $R \to M$ such that

$$X(\sigma + 2\pi) = gX(\sigma) \tag{49}$$

$(\mathcal{L}M)^g$ can be viewed as the space of sections of a certain M bundle

$$\begin{array}{ccc} M & \longrightarrow & W \\ & & \downarrow \\ & & S^1 \end{array} \tag{50}$$

over S^1.

It is easy to see that $(\mathcal{L}M)^g$ is a manifold, unlike —say —the naive quotient M/S^1 to which it is in some ways similar. In fact, the fiber bundle in (50) is trivial (but has no canonical trivialization), and as a manifold $(\mathcal{L}M)^g$ is isomorphic (but not canonically) to $\mathcal{L}M$.

Let us compare $(\mathcal{L}M)^g$ to $\mathcal{L}M$ from the standpoint of the group action that we are considering. Let P generate the rotation of S^1(thus$P \sim d/d\sigma$). Thus on $\mathcal{L}M$ or $(\mathcal{L}M)^g$, P generates the transformation $X(\sigma) \to X(\sigma + c)$, c being a constant. On $\mathcal{L}M$ it follows from (48) that

$$\exp 2\pi P = 1 \tag{51}$$

but on $(\mathcal{L}M)^g$ it follows from (49) that

$$\exp 2\pi P = g. \tag{52}$$

On $\mathcal{L}M$, P and K generate a group F isomorphic to $S^1 \times S^1$. On $(\mathcal{L}M)^g$, P and K generate a two dimensional abelian group F_g which is isomorphic to F but with no canonical choice of an isomorphism. We can therefore reasonably hope to learn something essentially new by studying equivariant index problems on $(\mathcal{L}M)^g$. Actually, it turns out that it is crucial to study the one parameter family $(\mathcal{L}M)^g$, $g\epsilon S^1$.

A formal expression for the equivariant index of the signature operator on $(\mathcal{L}M)^g$ could be written down as in section 2. However, we will not need this formula. It will suffice to observe the following. Let $g = e^{2\pi\alpha K}$, $0 \leq \alpha \leq 1$. Thus, α is an angular parameter on the S^1 group generated by K. At $\alpha = 0$, (51) asserts that the eigenvalues of P are integers. At $\alpha \neq 0$, this is not so. In an eigenspace of K (or more precisely of L_K) with eigenvalue k, (52) asserts that the eigenvalues of P are of the form $n + \alpha k$, n being an integer. Thus, the generalization of (45) to an equivariant index on $(\mathcal{L}M)^g$ at $g = \exp 2\pi\alpha K$ is of the general form

$$G_\alpha(q,\theta) = \Sigma\, c_{n,k}(\alpha)\, q^{n+\alpha k}\, e^{ik\theta}. \tag{53}$$

Now, in giving a sound mathematical basis for the supersymmetric nonlinear sigma model (even in a cutoff version), one would expect to prove that the signature operator on loop space has formal properties analogous to the properties of an elliptic operator in finite dimensions. In particular, the spectrum should vary smoothly with parameters such as α. This implies that the $c_{n,k}(\alpha)$ must be independent of α, since they arise in the solution of an equivariant index problem, and continuity of the spectrum implies that the index of an operator is invariant under continuous change of parameters.

Given that the $c_{n,k}(\alpha)$ are independent of α, we observe next that the index problem at $\alpha = 1$ is the same as the index problem at $\alpha = 0$. Setting $G_1(q,\theta) = G_0(q,\theta)$, we learn that

$$c_{n+k,k} = c_{n,k} \tag{54}$$

and as discussed earlier, this implies that $c_{n,k} = 0$ for $k \neq 0$.

The above may perhaps be clarified by the following considerations. On the infinite dimensional manifold $\mathcal{L}M$, if one works with the naive de Rham complex of differential forms of finite order, there is no analogue of Poincaré duality. The supersymmetric nonlinear sigma model corresponds rather to a theory in which one works near the "middle dimension" of the de Rham complex of $\mathcal{L}M$, with "semi-infinite" forms. The use of semi-infinite forms is implicit in the choice (24) of what we mean by the spin bundle of $\mathcal{L}M$. With the definitions as usually made in quantum field theory, there is in the supersymmetric nonlinear sigma model a transformation $*$ of the spin complex[*] which has the formal properties of Poincaré duality. In particular, the character-valued signature is

$$G(q,\theta) = Tr(*q^P e^{i\theta L_K}), \tag{55}$$

the trace running over the Hilbert space of harmonic sections of the signature complex —*i.e.*, zero energy states of the supersymmetric nonlinear sigma model.

★ To physicists it is the operator $(-1)^{F_L}$ that changes the sign of left-moving fermions but commutes with right movers.

Implicit in (55) is the statement that $G(q,\theta)$ exists as a function in some range of q and θ where the trace on the right-hand side converges. In our previous discussion, we treated $G(q,\theta)$ as a formal series. Indeed, one of the fundamental tasks in gaining mathematical understanding of the nonlinear sigma model is to show that the so called partition function converges absolutely for $|q| < 1$. As $G(q,\theta)$ is bounded above by the partition function, it should converge absolutely for $|q| < 1$, uniformly in θ.

In our previous discussion, we observed that the twisted loop space $(\mathcal{L}M)^g$ is isomorphic to $\mathcal{L}M$ but not canonically isomorphic to it. The obstruction to finding a natural identification of $(\mathcal{L}M)^g$ with $\mathcal{L}M$ really lies in the existence of a certain natural automorphism $\tau : \mathcal{L}M \rightarrow \mathcal{L}M$ which has been hidden in the previous discussion and which we will now make explicit. For a point $\gamma \epsilon \mathcal{L}M$, represented by a loop $\sigma \rightarrow X(\sigma)$, let $\tau\gamma$ be the loop

$$\sigma \longrightarrow (\exp \sigma L_K) X(\sigma) . \tag{56}$$

One readily sees that the transformation τ so defined obeys

$$\begin{aligned} \tau^{-1} L_K \tau &= L_K \\ \tau^{-1} P \tau &= P + L_K \end{aligned} \tag{57}$$

(The latter comes from the fact that P is essentially $d/d\sigma$.)

In finite dimensions, we would now pick a τ invariant metric in the definition of the Hodge dual operator $*$, and claim

$$\begin{aligned} G(q,\theta) &= Tr * q^P e^{i\theta L_K} \\ &= Tr * \tau^{-1} (q^P e^{i\theta L_K}) \tau \\ &= Tr * q^{P+L_K} e^{i\theta L_K} \\ &= G(q, \theta - i \ell n q) . \end{aligned} \tag{58}$$

(58) is in fact valid; it is easily seen to be equivalent to the relation (54) from which we originally extracted the fact that $c_{n,k} = 0$ for $k \neq 0$. However, in infinite dimensions the argument in (58) is too facile. Although we may well pick a K invariant metric on M, which will induce a τ invariant metric on $\mathcal{L}M$, the quantum field theory construction of "elliptic operators on $\mathcal{L}M$" involves additional terms, described in [1], which are not τ invariant. The supercharge D of the nonlinear sigma model (*i.e.* the Dirac operator in loop space) is transformed by τ into, say, $D^\tau = \tau^{-1} D \tau$. To justify (58), we must know that D and D^τ have the same equivariant index. This is so, since the family $D_x = xD + (1-x)D^\tau$, $0 \leq x \leq 1$, gives an equivariant interpolation from D to D^τ, and the D_x can be seen formally to be well-behaved operators on $\mathcal{L}M$. Consideration of the family D_x, with $D_{x=1}$ unitarily equivalent (via τ) to $D_{x=0}$, is an equivalent but perhaps illuminating way to re-express the spectral flow argument that we gave originally for our basic result on the $c_{n,k}$.

Together with the obvious fact $G(q,\theta) = G(q,\theta+2\pi)$, (58) states that for fixed q, $G(q,\theta)$ is a doubly periodic function on the complex plane with periods 2π and $-i\ell nq$. In other words, $G(q,\theta)$ is an elliptic function on the elliptic curve $\Sigma = C/\Lambda$, C being the complex plane and Λ the lattice generated by 2π and $-i\ell nq$. The doubly periodic character of $G(q,\theta)$ was first observed (and proved) by Ochanine, who defined $G(q,\theta)$ for real θ as a graded sum of character-valued indices of the operators D_{R_n}, and then showed (from the fixed point formula for the index of D_{R_n}) that the analytic continuation of $G(q,\theta)$ to complex θ is doubly periodic.

Actually, Feynman path integrals give an attractive conceptual explanation for the doubly periodic nature of $G(q,\theta)$. The "ordinary" equivariant signature $G(q,0)$ of loop space has a representation in terms of integrals over maps $\Sigma \to M$, Σ being the elliptic curve with periods 2π, $-i\ell nq$. The refinement $G(q,\theta)$ has an analogous representation in terms of integration over the space of sections of a twisted bundle,

$$\begin{array}{ccc} M & \longrightarrow & W \\ & & \downarrow \\ & & \Sigma \end{array} \tag{59}$$

The twisted bundles of interest are flat bundles, constructed by choosing a homomorphism $\pi_1(\Sigma) \to G$, G being a compact group that acts on M. With $G = S^1$, the choice of a homomorphism $\pi_1(\Sigma) \to G$ amounts to a choice of a point θ on the Jacobian of Σ. The function $G(q,\theta)$ that we have been discussing can be obtained as an integral over sections of the bundle (59), and its various properties (holomorphic in q and θ, doubly periodic in θ for fixed q, natural transformation law under a congruence subgroup of $SL(2,Z)$) follow from this representation by arguments that are fairly standard once one has developed a certain amount of quantum field theory machinery.

If one can show that the double periodic function $G(q,\theta)$ has no poles as a function of θ for fixed q, then, since an elliptic function without poles must be constant, $G(q,\theta)$ must be independent of θ. That $G(q,\theta)$ is independent of θ is precisely our desired result $c_{n,k} = 0$, $k \neq 0$. From the standpoint of quantum field theory, the absence of poles in $G(q,\theta)$, as a function of θ for fixed q, follows from the fact that the series defining $G(q,\theta)$ is absolutely convergent for $|q| < 1$, being dominated by the partition functions of supersymmetric nonlinear sigma models on the twisted loop spaces $(\mathcal{L}M)^g$. This reasoning might give a somewhat different approach to using quantum field theory to prove the result about the $c_{n,k}$.

We have here concentrated on the signature operator on $\mathcal{L}M$ because this is in many ways the most canonical case in which the anomaly criterion of section 3 is obeyed. However, if V is any vector bundle obeying the criterion $p_1(V_G) = p_1(T_G)$ discussed in section 3, then the arguments that we have given generalize immediately to the "Dirac operator with values in V" and the "elliptic Euler characteristic of V"; the relevant formulas were given in equations (31) and (34). In each case one gets $c_{n,k} = 0$ for $k \neq 0$.

5. Complex Manifolds

So far in this paper, we have been working in the context of real differential geometry. However, the supersymmetric nonlinear sigma model has particularly rich properties for special classes of manifolds —almost complex manifolds, Kahler manifolds, and hyper-Kahler manifolds. It is natural to expect that the theorem on equivariant characteristic numbers that we discussed in the last section would have analogues for those special classes of manifolds. The purpose of this section is to make a start in this direction. In particular, we will sketch a generalization to elliptic genera of certain results of Hattori [16].

To begin with, let M be a manifold of dimension d, and let W be a complex vector bundle over M of complex dimension k, endowed with a hermitian metric. Let $\overline{W}$ be the dual of W. Then $W \oplus \overline{W} = V_C$ is naturally the complexification of a $2k$ dimensional real vector bundle V. Let L be the canonical line bundle $L = \Lambda^k W$. We will be interested in a situation in which for some integer $N > 1$, L has an Nth root ξ. In other words, we will assume the existence of a line bundle ξ with $\xi^N = L$. We will describe constructions that associate to this data modular forms for the subgroups $\Gamma_0(N)$ of $SL(2,Z)$. For $N = 2$ the constructions do not require the complex structure of V and reduce to those of section 2.

A section of V is a pair $(f, \bar{f})$, with f a section of W and $\bar{f}$ its complex conjugate. If ζ is a complex number of modulus one, then $(\zeta f, \bar{\zeta}\bar{f})$ is likewise a section of V; this gives an action of the group H of complex numbers of modulus one on sections of V.

As in section 2, let $\gamma : S^1 \to M$ be a loop in M, that is, a point in $\mathcal{L}M$. Let $\gamma^* V$ be the pullback of V from M to S^1 via γ, and let V_γ be the space of sections of $\gamma^* V$. Then H acts on V_γ in an obvious way, *i.e.* $(f(\sigma), \bar{f}(\sigma)) \to (\zeta f(\sigma), \zeta^{-1}\bar{f}(\sigma))$. As γ varies, the V_γ are fibers of a vector bundle $\mathcal{V}$ over $\mathcal{L}M$. We already described in equation (31) the spinor bundle $\Delta(\mathcal{V})$ restricted to $M \subset \mathcal{L}M$:

$$\Delta(\mathcal{V})|_M = \Delta(V) \bigotimes_{m=1}^{\infty} \Lambda_{q^m} V \tag{60}$$

We now wish to refine this to keep track of the action of complex numbers of modulus one. We write symbolically $V = \zeta W \oplus \zeta^{-1}\overline{W}$, and observe that

$$\Lambda_{q^m} V = \Lambda_{q^m}(\zeta W \oplus \zeta^{-1}\overline{W}) = \Lambda_{q^m}(\zeta W) \otimes \Lambda_{q^m}(\zeta^{-1}\overline{W}). \tag{61}$$

Also, from standard facts about the spin representation, it follows that

$$\begin{aligned} \Delta(V) &= \Delta(\zeta W \oplus \zeta^{-1}\overline{W}) \\ &= \zeta^{k/2} L^{1/2} \Lambda(\zeta^{-1}\overline{W}) \end{aligned} \tag{62}$$

with $L^{1/2}$ a square root of L. The refinement of (60) to keep track of the action of H is thus

$$\Delta(\mathcal{V})|_M = \zeta^{k/2}L^{1/2}\Lambda(\zeta^{-1}\overline{W})\bigotimes_{m=1}^{\infty}\Lambda_{q^m}(\zeta W)\bigotimes_{r=1}^{\infty}\Lambda_{q^r}(\zeta^{-1}\overline{W}). \tag{63}$$

Now, we consider the Dirac index for spinors on $\mathcal{L}M$ twisted by $\Delta(\mathcal{V})$. More precisely, we wish to consider the equivariant Dirac index, equivariant under the natural circle action on $\mathcal{L}M$ and the action of H. The generalization of equation (31) is simply

$$\begin{aligned} F_V(q,\zeta) = q^{-\frac{(d-2k)}{24}}\Big\langle \hat{A}(M)\mathrm{ch}\bigotimes_{\ell=1}^{\infty}S_{q^\ell}T\otimes\zeta^{k/2}L^{1/2}\Lambda(\zeta^{-1}\overline{W}) \\ \bigotimes_{m=1}^{\infty}\Lambda_{q^m}(\zeta W)\bigotimes_{r=1}^{\infty}\Lambda_{q^r}(\zeta^{-1}\overline{W}),\, M\Big\rangle \end{aligned} \tag{64}$$

For (64) to be the equivariant index of an operator on $\mathcal{L}M$, it is necessary to impose certain restrictions on characteristic classes, which in our previous discussion were $w_2(V) = w_2(T), p_1(V) = p_1(T)$. The main novelty is that the requirement $p_1(T) = p_1(V)$ must be imposed equivariantly with respect to the action of H. (T is the tangent bundle of M.) This condition is in fact obeyed for the whole group $H \cong U(1)$ if and only if L is trivial. If, however, L is not trivial, but is the Nth power of a line bundle ξ, then (64) is an equivariant index provided that $(-\zeta)^N = 1$. In this case, (64) is a modular form for a congruence subgroup of $SL(2,Z)$ conjugate to $\Gamma_1(N)$. For special cases, the restriction $(-\zeta)^N = 1$ and the statement that (64) gives modular forms of $\Gamma_0(N)$ have already entered in section 2. For $N = 1$, $\zeta = -1$, this is the statement that the "Euler characteristic of a vector-bundle" given in equation (34) is a modular form for $\Gamma_1(1) = SL(2,Z)$. For $N = 2$, $\zeta = 1$, this is the statement that the "signature of a vector bundle" (31) is a modular form of level two. The physical origin of the restriction on ζ is that if $-\zeta$ is of order N, then instanton amplitudes are invariant under the twist by ζ only if the first Chern class of L is divisible by N or in other words only if L has an Nth root. I will not attempt here to elucidate the connection of this statement with the equivariant p_1.

Since (64) leads to modular forms of level N, by making transformations by $SL(2,Z)$ matrices that are not in $\Gamma_1(N)$, (64) can be related to various other and essentially equivalent formulas. Let $-\zeta$ be a primitive N^{th} root of 1, and let s be any positive integer in the range $1 \leq s \leq N-1$ which is relatively prime to N. Then a suitable $SL(2,Z)$ transformation will transform (64) into the form

$$\begin{aligned} \tilde{F}_V(q,s) = q^{-\frac{d}{24}-\frac{(1-3(1-2s/N)^2)k}{24}}\Big\langle \hat{A}\,\mathrm{ch}\bigotimes_{n=1}^{\infty}S_{q^n}T \\ \bigotimes L^{-\frac{1}{2}+\frac{s}{N}}\bigotimes_{m=0}^{\infty}\Lambda_{-q^{m+s/N}}W\bigotimes_{m=0}^{\infty}\Lambda_{-q^{m+1-s/N}}\overline{W},\, M\Big\rangle. \end{aligned} \tag{65}$$

(65) is the equivariant index of a Dirac operator on $\mathcal{L}M$ twisted by a certain generalization of the bundle $\Delta(\hat{T})$ of equation (30). (Essentially, one uses a complex line bundle over S^1 of order N, a

notion that makes sense equivariantly with respect to the rotation of S^1 , in place of the Hopf bundle ϵ that was used in constructing $\Delta(\hat{T})$.) In any case, since (64) and (65) are related by $SL(2,Z)$, any true statement about (64) is equivalent to a true statement about (65).

Now, we assume that M admits the action of a compact connected Lie group G, which we may as well take to be $G = S^1$. Suppose that the G action lifts to an action on W and that in the G-equivariant sense $p_1(T_G) = p_1(V_G)$. Then we can use the arguments of section 4 to show that the G-index generalization of (64) or (65) to a character of G gives in fact a multiple of the trivial character.

A natural way to obey $p_1(T_G) = p_1(V_G)$ is to suppose that M is an almost complex manifold, and pick $V = T$, with the G action on T and V induced from that on M.* In this case, the statement that the G-index generalization of (64) and (65) is a multiple of the trivial G-character gives an interesting restriction on almost complex manifolds. For instance, if we look at (65) and specialize to the lowest power of q, we learn that the G-index of spinors on M twisted by $L^{-\frac{1}{2}+\frac{s}{N}}$, or in other words the G-equivariant extension of

$$\left\langle \hat{A}(M) \operatorname{ch} L^{-\frac{1}{2}+\frac{s}{N}}, M \right\rangle \tag{66}$$

is a multiple of the trivial G character for $s = 1, 2, \cdots, N-1$. Hattori actually proved that this vanishes. Our results are thus analogues of his for the higher powers of q.

$\star$ More generally, M may be stably almost complex, and then we take $V = T \oplus \theta \oplus \cdots \oplus \theta$, with θ a trivial line bundle. In this case the lifting of G to act on V may be any such that $p_1(V_G) = p_1(T_G)$.

REFERENCES

[1] E. Witten, *J. Diff. Geom.* **17** (1982) 661; and in *Anomalies, Geometry, and Topology*, ed. W. Bardeen and A. White (World Scientific, 1985), especially pp. 91-95.

[2] A. Schellekens and N. Warner, *Phys. Lett.* **177B** (1986) 317, "Anomaly Cancellation and Self-Dual Lattices" (MIT preprint, 1986), "Anomalies, Characters, and Strings" (CERN preprint TH-4529/86); K. Pilch, A. Schellekens, and N. Warner (preprint, 1986).

[3] E. Witten, "Elliptic Genera and Quantum Field Theory" (Princeton preprint PUPT-1024, 1986), to appear in *Comm. Math. Phys.*

[4] A. Pressley and G.B. Segal, *Loop Groups and Their Representations* (Oxford University Press, 1986).

[5] M.F. Atiyah and I. Singer, *Ann. Math.* **87** (1968) 484, 586;

M.F. Atiyah and G.B. Segal, *Ann. Math.* **87** (1968) 531.

[6] E. Witten, "Physics and Geometry," address at the International Congress of Mathematicians, Berkeley, August 1986, section IV.

[7] G. Moore and P. Nelson, *Phys. Rev. Lett.* **53** (1984) 1519.

[8] T. Killingback, "World Sheet Anomalies and Loop Geometry," Princeton preprint (1986).

[9] I.B. Frenkel, "Spinor Representations of Affine Lie Algebras," *Proc. Nat. Acad. Sci. USA* **77** (1980) 6303; "Two Constructions of Affine Lie Algebra Representations," *J. Funct. Anal.* **44** (1981) 259.

[10] M.F. Atiyah and F. Hirzebruch, in *Essays in Topology and Related Subjects* (Springer-Verlag, 1970), p. 18.

[11] E. Witten, "Fermion Quantum Numbers in Kaluza-Klein Theory," in the *Proceedings of the 1983 Shelter Island Conference on Quantum Field Theory and the Foundations of Physics*, eds. N. Khuri et al. (MIT Press, 1985).

[12] P. Landweber and R.E. Stong, "Circle Actions on Spin Manifolds and Characteristic Numbers," Rutgers preprint (1985), to appear in *Topology*;

P. Landweber, "Elliptic Cohomology and Modular Forms," to appear in this volume.

[13] S. Ochanine, "Sur les genres multiplicatifs définis par des intégrales elliptiques," to appear in *Topology*; "Genres Elliptiques Equivariantes," to appear in this volume.

[14] D. V. and G. V. Chudnovsky, "Elliptic Modular Forms and Elliptic Genera," Columbia University preprint (1985), to appear in *Topology*.

[15] D. Zagier, "Note on the Landweber-Stong Elliptic Genus," to appear in this volume.

[16] A. Hattori, "Spin Structures and S^1 Actions," *Inv. Math.* **48** (1978) 7.

JACOBI QUARTICS, LEGENDRE POLYNOMIALS AND FORMAL GROUPS

by Noriko Yui*

Department of Mathematics
University of Toronto
Toronto, Ontario
Canada M5S 1A1

Introduction

Let J_ρ denote the Jacobi quartic (of modulus ρ) defined inhomogenously by the equation

$$(*) \qquad Y^2 = 1 - 2\rho X^2 + X^4$$

over the function field $K = \mathbb{Q}(\rho)$ where ρ is a transcendental variable over the field $\mathbb{Q}$ of rational numbers. There is associated to J_ρ the non-zero differential 1-form of the first kind

$$\omega_\rho = \frac{dX}{Y} = \frac{dX}{\sqrt{1 - 2\rho X^2 + X^4}} = \sum_{n=0}^{\infty} P_n(\rho)\, X^{2n}\, dX$$

where $P_n(\rho)$ denotes the n-th Legendre polynomial.

Now choose an odd prime p, and fix it once and for all. We can extend the p-adic valuation ord_p of $\mathbb{Q}$ (normalized so that $\mathrm{ord}_p(p) = 1$) to $K = \mathbb{Q}(\rho)$ by the Gaussian definition, denoting it ν_p. The corresponding valuation ring, R, is an unramified discrete

---

* Noriko Yui was partially supported by the Natural Sciences and Engineering Research Council of Canada (NSERC) under grant # A8566.

valuation ring with the maximal ideal pR and the residue field $\mathbb{F}_p(\rho)$ where $\mathbb{F}_p$ is the prime field of p elements. Moreover, the field $K = \mathbb{Q}(\rho)$ endowed with the valuation ν_p satisfies the following condition

$$(F_1)\quad \left[\begin{array}{l}\text{The valuation } \nu_p \text{ of } K \text{ is unramified,}\\ \text{and there is an endomorphism } \sigma : K \to K\\ \text{such that } \alpha^\sigma \equiv \alpha^p \pmod{pR} \text{ for any}\\ \alpha \in R.\end{array}\right]$$

Under this circumstance, the theory of commutative formal groups due to Honda [3] is at our disposal. The purpose of this paper is to study the Jacobi quartic J_ρ, Legendre polynomials $P_n(\rho)$ $(n \in \mathbb{N})$ and the formal groups associated to them in the framework of the Honda theory of commutative formal groups.

It is classically known that Legendre polynomials satisfy the Schur congruence : If $n = a_0 + a_1 p + a_2 p^2 + \ldots + a_\mu p^\mu \in \mathbb{N}$ where $a_i < p$ for every i, then

$$P_n(\rho) \equiv P_{a_0}(\rho)\, P_{a_1}(\rho^p)\, P_{a_2}(\rho^{p^2}) \cdots P_{a_\mu}(\rho^{p^\mu}) \qquad \pmod{pR}.$$

The Schur congruence for $P_{(p^\mu-1)/2}(\rho)$ is reformulated in the context of the Honda theory (Proposition (II.2.1)). The Schur congruence appears to be tied up with some arithmetical and geometrical properties of the Jacobi quartic J_ρ. For instance, the Schur congruence for $P_{(p^\mu-1)/2}(\rho)$ is equivalent to the uniqueness of the representation of the iterated Cartier operator $\mathcal{C}^\mu$ on $H^0(\bar{J}_\rho, \Omega^1)$ (Proposition (II.1.3)), and also to the p-integrality of the formal group law F_{J_ρ} of J_ρ (Theorem (II.2.5)).

The structure of the weak isomorphism class of the formal group law F_{J_ρ} of J_ρ is determined : the class of F_{J_ρ} is parametrized over the completion $\hat{R}$ of R by the Hasse invariant $H(\bar{J}_\rho)$ of $\bar{J}_\rho$, and hence by the $(p-1)/2$-th Legendre polynomial $P_{(p-1)/2}(\rho)$ (Theorem (II.3.2)).

The Jacobi quartic J_ρ and the holomorphic differential 1-form ω_ρ are studied under a different parametrization. Let $t = X/Y$ be a local parameter at zero of J_ρ. Then

$$\omega_\rho = \sum_{n=0}^{\infty} a_{2n}(\rho)\, t^{2n}\, dt$$

with $a_0(\rho) = 1$ and $a_{2n}(\rho) \in \mathbb{Z}[2\rho] \subset R$ (Proposition (I.6.2)). The fact that the iterated Cartier operator $\mathcal{C}^\mu$ has a unique representation on $H^0(\bar{J}_\rho, \Omega^1)$ independently of the choice of a local parameter at zero of J_ρ gives rise to the congruence

$$a_{p^\mu-1}(\rho) \equiv P_{(p^\mu-1)/2}(\rho) \pmod{pR} \quad \text{for any} \quad \mu \geq 0$$

(Proposition (II.4.3)). Furthermore, in this parametrization, we have the following congruence

$$a_{p^\mu-1}(\rho) \equiv a_{p-1}(\rho)\, a_{p-1}(\rho^p)\, a_{p-1}(\rho^{p^2}) \cdots a_{p-1}(\rho^{p^\mu}) \pmod{pR}$$

which we may call the "Atkin and Swinnerton-Dyer" type congruence for J_ρ modulo pR (Theorem (II.4.4)). The formal group law of J_ρ is also discussed under this parametrization (II.5).

There are the Honda congruences for Legendre polynomials modulo higher powers of p :

$$P_{mp^{\mu+1}}(\rho) \equiv P_{mp^\mu}(\rho^p) \pmod{p^{\mu+1}R}$$

and

$$P_{mp^{\mu+1}-1}(\rho) \equiv P_{mp^\mu-1}(\rho^p) \pmod{p^{\mu+1}R}$$

for every $\mu \geq 0$. These congruences reflect some arithmetical and geometrical properties of the rational curve

$$C_\rho \;:\; Y^2 = 1 - 2\rho X + X^2$$

defined over $K = \mathbb{Q}(\rho)$. The Jacobi quartic J_ρ is a double cover of C_ρ over K. The Honda congruences are equivalent to the fact that the isomorphism class of the formal group law of C_ρ is represented by the multiplicative formal group law $\hat{\mathbb{G}}_m^{\pm}$ (Proposition (III.2.1)). Relations between the second Honda congruence and the Schur congruence are discussed (Remark (III.1.3)(b)).

Finally, in the appendix, we discuss the formal group associated to the twisted Legendre polynomials $P^{\sigma^{\mu}-1}_{(p-1)/2}(\rho)$ with $\mu \geq 0$.

Notations and Conventions

The symbol $\binom{a}{b}$ denotes the binomial coefficient. For a set S, $\# S$ stands for the cardinality of S. We write $R[x_1,\ldots,x_n]$ (resp. $R[[x_1,\ldots,x_n]]$) for the ring of polynomials (resp. the ring of formal power series) in the variables $x_1,\ldots,x_n$ over a ring R. For f and g in $R[[x]]$, write $f \equiv g \pmod{\deg r}$ if $f - g$ contains no monomials of degree less than r, and $f \equiv g \pmod{p^{\mu}R}$ $(\mu \in \mathbb{N})$ if the coefficient of x^i in $f(x) - g(x)$ is divisible by p^{μ} for every $i \geq 0$.

All formal groups discussed in this paper are commutative and 1-dimensional. We simply use the terminology "formal group" for a commutative 1-dimensional formal group. $\hat{\mathbb{G}}_m^{\pm}$ and $\hat{\mathbb{G}}_a$ denote, respectively, the multiplicative formal group law and the additive formal group law.

Two formal groups F and G over a ring R are said to be <u>weakly isomorphic over</u> R if there is an element $\phi(x) \in R[[x]]$ such that $\phi(x) \equiv Cx \pmod{\deg 2}$ with C a unit in R, and that $\phi \circ F = G \circ \phi$. Such a ϕ is called a <u>weak isomorphism</u>. Furthermore, a weak isomorphism ϕ is called an <u>isomorphism</u> if $\phi(x) \equiv x \pmod{\deg 2}$. We denote by $\mathrm{Iso}_R(F,G)$ the set of all weak isomorphisms from F to G over R. The notion of weak isomorphism (resp. isomorphism) defines an equivalence relation in the set of formal groups over R, and the equivalence class of a formal group F is called the <u>weak isomorphism</u> (resp. <u>isomorphism</u>) <u>class of</u> F <u>over</u> R.

Acknowledgements

I am indebted to Professors Gerhardt Frey and Christian U. Jensen for their helpful and stimulating discussions on this subject.

Thanks are also due to Professors David and Gregory Chudnovsky, and Professor Peter Landweber for their interest in my work on this topic.

I. Jacobi quartics

I.1. The Jacobi quartic (References : Igusa [5,6]).

Consider a plane curve defined inhomogenously by the equation (*) over the function field $K = \mathbb{Q}(\rho)$ where ρ is a transcendental variable over $\mathbb{Q}$. The curve (*) is absolutely irreducible, and has its only singularity at the point at infinity $(0,1,0)$. This curve is of genus 1. Therefore, we can introduce a group law of composition over K with the point $(0,1)$ as its zero element. The curve (*) with this group law is called the <u>Jacobi quartic (of modulus</u> ρ) and denoted J_ρ.

Resolving the singularity $(0,1,0)$ on J_ρ, we obtain a nonsingular model, A_ρ, which is an elliptic curve (a complete group variety of dimension 1) over K. In fact, the morphism $A_\rho \to J_\rho$ is everywhere biregular except at the singular point. If $u \in A_\rho$, then we denote by $(x,y) = (x(u), y(u))$ the corresponding point on J_ρ by the map $A_\rho \to J_\rho$.

The discriminant, Δ_ρ, and the absolute invariant, j_ρ, of J_ρ are given by

(I.1.1) $$\Delta_\rho = 2^8(1-\rho^2)^2 \quad \text{and} \quad j_\rho = 2^6 \frac{(3+\rho^2)^3}{(1-\rho^2)^2} .$$

Associated to J_ρ, there is the non-zero differential 1-form of the first kind

(I.1.2) $$\omega_\rho = \frac{dX}{Y} = \frac{dX}{\sqrt{1-2\rho X^2 + X^4}} .$$

ω_ρ gives a basis of the 1-dimensional K-vector space $H^0(J_\rho, \Omega^1) \cong H^0(A_\rho, \Omega^1)$ of holomorphic differential 1-forms on J_ρ. It has a formal power series expansion with respect to X which is given as follows:

(I.1.3) $$\omega_\rho = \sum_{n=0}^{\infty} P_n(\rho)\, X^{2n}\, dX$$

where $P_n(\rho)$ is the n-th Legendre polynomial defined as

$$(\text{I.1.4})\qquad \begin{cases} P_0(\rho) = 1 \\ P_n(\rho) = \sum_{r=0}^{m} (-1)^r \dfrac{(2n-2r)!}{2^n\, r!(n-r)!(n-2r)!} \rho^{n-2r} \end{cases}$$

with $m = n/2$ or $(n-1)/2$ whichever is an integer (cf. Whittaker and Watson [11]). $P_n(\rho)$ is a polynomial in ρ of degree n over $\mathbb{Z}[1/2]$.

I.2. The reduced Jacobi quartic.

Let p be a rational prime different from 2. We can extend the p-adic valuation ord_p of $\mathbb{Q}$ (normalized by $\mathrm{ord}_p(p) = 1$) to $K = \mathbb{Q}(\rho)$ by the Gaussian definition, denoting it ν_p. In fact, for $f = a_0 + a_1\rho + a_2\rho^2 + \ldots + a_n\rho^n \in \mathbb{Q}[\rho]$, let

$$\nu_p(f) := \min_{0 \le i \le n} \{ \mathrm{ord}_p(a_i) \}$$

and for $f/g \in K$ with $f, g \in \mathbb{Q}[\rho]$, $g \neq 0$, let

$$\nu_p(f/g) = \nu_p(f) - \nu_p(g).$$

Let R denote the corresponding valuation ring of K. Then it is easy to see that R is an unramified discrete valuation ring with the maximal ideal pR and the residue field $\mathbb{F}_p(\rho)$. R contains rings $\mathbb{Z}[1/2][\rho]$ and $\mathbb{Z}[1/2][1/\rho]$. Let $\overline{K}$ (resp. $\overline{R}$) denote the algebraic closure of K (resp. the integral closure of R in $\overline{K}$).

The Jacobi quartic J_ρ of modulus ρ is clearly integral with respect to ν_p. We see from (I.1.1) that $\nu_p(\Delta_\rho) = 0$ and $\nu_p(j_\rho) = 0$. This means that J_ρ has good reduction at ν_p and the reduction, $\overline{J}_\rho := J_\rho \pmod{pR}$, defines a Jacobi quartic (of modulus ρ) over $\mathbb{F}_p(\rho)$, which we call the <u>reduced Jacobi quartic (of modulus ρ)</u>.

I.3. The group law on the Jacobi quartic.

The Jacobi quartic J_ρ has a group law of composition over K with the point $(0,1)$ as its zero element. This group law was found already by Euler (cf. Igusa [5] and Landweber [8]).

(I.3.1) The Addition Theorem on J_ρ. Let $u, v \in A_\rho$ and let $(x(u), y(u))$ and $(x(v), y(v))$ denote the corresponding points on J_ρ by the map $A_\rho \to J_\rho$. Then

$$x(u+v) = \frac{x(u)y(v) + x(v)y(u)}{1 - x(u)^2x(v)^2}$$

$$= \frac{x(u)\sqrt{1-2\rho x(v)^2+x(v)^4} + x(v)\sqrt{1-2\rho x(u)^2+x(u)^4}}{1 - x(u)^2x(v)^2} .$$

In particular, for $u = v$, we have

$$x(2u) = \frac{2x(u)y(u)}{1-x(u)^4} = \frac{2x(u)\sqrt{1-2\rho x(u)^2+x(u)^4}}{1 - x(u)^4} .$$

The Addition Theorem on J_ρ can be described in terms of formal group law on J_ρ as follows.

(I.3.2) Proposition. The formal group law on J_ρ is given by a formal power series F_{J_ρ} in two variables X_1 and X_2 over the ring $\mathbb{Z}[1/2][\rho] \subset R$, and it starts off

$$\begin{aligned} F_{J_\rho}(X_1,X_2) &= X_1 + X_2 - \rho(X_1^2X_2 + X_1X_2^2) + (X_1^3X_2^2 + X_1^2X_2^3) \\ &+ (\tfrac{1}{2} - \tfrac{1}{2}\rho^2)(X_1^4X_2 + X_1X_2^4) + (\tfrac{1}{2}\rho - \tfrac{1}{2}\rho^3)(X_1^6X_2 + X_1X_2^6) \\ &- \rho(X_1^4X_2^3 + X_1^3X_2^4) + (-\tfrac{1}{8} + \tfrac{3}{4}\rho^2 - \tfrac{5}{8}\rho^4)(X_1^8X_2 + X_1X_2^8) \\ &+ (\tfrac{1}{2} - \tfrac{1}{2}\rho^2)(X_1^6X_2^3 + X_1^3X_2^6) + (X_1^5X_2^4 + X_1^4X_2^5) + \cdots . \end{aligned}$$

Proof. Put $X_1 = x(u)$ and $X_2 = x(v)$, and expand out $\sqrt{1-2\rho X_1^2+X_1^4}$ and $\sqrt{1-2\rho X_2^2+X_2^4}$ into Taylor series at $X_1 = 0$ and $X_2 = 0$, respectively. Expand also $(1-X_1^2X_2^2)^{-1}$ into a Taylor series at $X_1 = X_2 = 0$. It is immediately seen that F_{J_ρ} is a formal power series with all the coefficients in $\mathbb{Z}[1/2][\rho]$. That F_{J_ρ} is a formal group law follows from Theorem (I.3.1).

I.4. The division polynomials of the Jacobi quartic.

If m is an odd natural number, the multiplication by m on A_ρ is an endomorphism of degree m^2, and it induces the transformation of J_ρ, called a multiplication by m on J_ρ. If $u \in A_\rho$, $(x(mu), y(mu))$ denotes the point on J_ρ corresponding to the image $mu \in A_\rho$ of u under the endomorphism m.

(I.4.1) Theorem (Kronecker ; Igusa [5]). There exist two even polynomials $F_m(X)$ and $G_m(X)$ with coefficients in $\mathbb{Z}[8\rho] \subset R$ for which

$$x(mu) = (-1)^{(m-1)/2} x^{m^2} F_m(x^{-1}) F_m(x)^{-1}$$

$$y(mu) = y\, G_m(x) F_m(x)^{-2}$$

with $x = x(u)$ and $y = y(u)$.

More precisely, $F_m(X)$ and $G_m(X)$ are given as follows:

$$F_m(X) = \prod_{\substack{u \in A_\rho \\ mu = 0}} (1 - x(u)X) = 1 + \sum_{0 \leq 2i < m-1} c_i X^{m^2-1-2i}$$

with $\deg F_m = m^2 - 1$, and

$$G_m(X) = \prod_{\substack{s \bmod 2r \\ u \in H}} (X - x(u+s))$$

with $\deg G_m = 2m^2 - 2$. Here r denotes a primitive 4-th division point of A_ρ, and H a half system of the abelian group $\mathrm{Ker}\,(m : A_\rho \to A_\rho)$. (Cf. Igusa [5], p. 441 for the definition of a half system of an abelian group.)

(I.4.2) Definition-Proposition (cf. Igusa [5]). With the notations of Theorem (I.4.1) in force, put

$$T_m(X) = X^{m^2} F_m(X^{-1}).$$

Then

$$T_m(X) = \prod_{\substack{u \in A_\rho \\ mu = 0}} (X - x(u))$$

and it is a polynomial in X of degree m^2, which is called the m-th division polynomial of the Jacobi quartic J_ρ.

In particular, if $m = p$ is an odd prime, then

$$T_p(X) = X^p \{ (X^p)^{p-1} + \sum_{0<2i<p-1} \gamma_i(\rho) P_{(p-1)/2}(\rho) (X^p)^{2i}$$

$$+ (-1)^{(p-1)/2} P_{(p-1)/2}(\rho) \}$$

where $P_{(p-1)/2}(\rho)$ is the $(p-1)/2$-th Legendre polynomial and $\gamma_i(\rho)$ is a polynomial in ρ over R for all i, $0 < 2i < p-1$.

Furthermore,

$$\prod_{\substack{u \in A_\rho \\ pu = 0}} x(u) = (-1)^{(p-1)/2} P_{(p-1)/2}(\rho) = \left(\frac{-1}{p}\right) P_{(p-1)/2}(\rho).$$

(Here $(-1/p)$ is the Legendre symbol.)

Let m be an odd natural number. Then the endomorphism $m : A_\rho \to A_\rho$ induces the K-linear map m^* on $H^0(A_\rho, \Omega^1)$ and hence on $H^0(J_\rho, \Omega^1)$. We compute the effect of m^* on a K-basis ω_ρ of $H^0(J_\rho, \Omega^1)$. We have, on the one hand,

$$m^*(\omega_\rho) = m\,\omega_\rho = m \sum_{n=0}^{\infty} P_n(\rho) X^{2n}\, dX,$$

and on the other hand,

$$m^*(\omega_\rho) = m^*\left(\frac{dx(u)}{y(u)}\right) = \frac{dx(mu)}{y(mu)}$$

$$= \frac{(-1)^{(m-1)/2}\, d\{x^{m^2} F_m(x^{-1}) F_m(x)^{-1}\}}{y G_m(x) F_m(x)^{-2}} \qquad (x = x(u)).$$

In particular, if $m = p$ is an odd prime, then we have

$$p^*(\omega_\rho) = (-1)^{(p-1)/2} \frac{F_p(X)\,d(T_p(X)) - T_p(X)\,d(F_p(X))}{Y\,G_p(X)}$$

$$= p \sum_{n=0}^{\infty} P_n(\rho)\,X^{2n}\,dX.$$

Comparing the coefficient of X^{p-1} in both expressions, we obtain the following result, which is in concordance with (I.4.2).

(I.4.3) Proposition.

$$P_{(p-1)/2}(\rho) = \left(\frac{-1}{p}\right)(\text{the coefficient of } X^p \text{ in } T_p(X)).$$

For a prime $p > 2$, we now consider the endomorphism $[p] := [p]_{F_{J_\rho}}$ (the multiplication by p in the formal group law F_{J_ρ} of J_ρ). Using the formal power series representation of Proposition (I.3.2), we get

$$[1](X) = X,$$

$$[2](X) = F_{J_\rho}(X,X) = 2X - 2\rho X^3 + 2X^4 + (1-\rho^2)X^5 + \cdots ,$$

$$[3](X) = F_{J_\rho}([2](X), X) = 3X - 8\rho X^3 + \ldots$$

$$\equiv \rho X^3 \equiv P_1(\rho)X^3 \quad (\text{mod } 3R, \text{ mod deg } 4),$$

and inductively,

$$[p](X) = F_{J_\rho}([p-1](X), X).$$

(I.4.4) Proposition. Let p be an odd prime, and let $[p]$ denote the multiplication by p endomorphism in the formal group law F_{J_ρ}. Then

$$[p](X) \equiv P_{(p-1)/2}(\rho)\,X^p \quad (\text{mod } pR, \text{ mod deg } p+1)$$

Furthermore, $[p](X)$ is a power series in X^p. Therefore, the reduced formal group $\overline{F}_{J_\rho} =: F_{J_\rho}$ (mod pR) has height $h = 1$.

Proof. Let pl_{J_ρ} (resp. $pl_{\overline{J}_\rho}$) denote the multiplication by p on J_ρ (resp. $\overline{J}_\rho$). Let $J_\rho(\overline{K})$ denote the group of points on J_ρ

defined over the algebraic closure $\bar{K}$ of K. Then we have

$$\operatorname{Ker} pl_{J_\rho} = \operatorname{Ker} pl_{\bar{J}_\rho} \oplus \operatorname{Ker}[p]$$

where

$$\operatorname{Ker} pl_{J_\rho} = \{ x = x(u) \in J_\rho(\bar{K}) \mid T_p(x) = 0 \} \quad \text{with} \quad \# \operatorname{Ker} pl_{J_\rho} = p^2,$$

and

$$\# \operatorname{Ker} pl_{\bar{J}_\rho} = \text{the separable degree of } pl_{\bar{J}_\rho} : \bar{J}_\rho \to \bar{J}_\rho = p,$$

$$\# \operatorname{Ker}[p] = \text{the inseparable degree of } pl_{\bar{J}_\rho} : \bar{J}_\rho \to \bar{J}_\rho = p.$$

(As $T_p(X)$ is a polynomial in X^p, we may conclude that $pl_{\bar{J}_\rho}$ has inseparable degree p, and hence $\# \operatorname{Ker}[p] = p$. Since

$$\prod_{\substack{u \in A_\rho \\ pu=0}} x(u) = (-1)^{(p-1)/2} P_{(p-1)/2}(\rho) \neq 0,$$

the group, ${}_p\bar{J}_\rho(\bar{K})$, of p-torsion points on $\bar{J}_\rho$ has the structure ${}_p\bar{J}_\rho(\bar{K}) \cong \mathbb{Z}/p\mathbb{Z}$. Hence $\bar{J}_\rho$ is "ordinary", and accordingly, $pl_{\bar{J}_\rho}$ has separable degree p, and hence $\# \operatorname{Ker} pl_{\bar{J}_\rho} = p$.) Now by Theorem (I.4.1), we have

$$[p](X) = (-1)^{(p-1)/2} T_p(X)/F_p(X)$$

and $T_p(X)$ and $F_p(X)$ are polynomials in X^p by (I.4.2). Thus $[p](X)$ is a power series in X^p, and the required congruence follows from this identity reading it modulo pR and modulo deg $p+1$. As $P_{(p-1)/2}(\rho) \neq 0$, the formal group law $F_{\bar{J}_\rho}$ is of height $h = 1$.

I.5. The Hasse invariant $H(\bar{J}_\rho)$ of the reduced Jacobi quartic.

We consider the reduced Jacobi quartic $\bar{J}_\rho$ defined over $\mathbb{F}_p(\rho)$. Let $\bar{\omega}_\rho$ be the differential of the first kind on $\bar{J}_\rho$. It has the formal power series expansion (I.1.3) with all the coefficients reduced modulo pR, and it gives a basis of the 1-dimensional vector space $H^0(\bar{J}_\rho, \Omega^1)$ of holomorphic differential 1-forms on $\bar{J}_\rho$ over $\mathbb{F}_p(\rho)$. Considering both $\bar{J}_\rho$ and $\bar{\omega}_\rho$ over the perfect closure $\mathbb{F}_p(\rho, \rho^{1/p^\infty}) := \bigcup_{n \geq 0} \mathbb{F}_p(\rho, \rho^{1/p^n})$ of $\mathbb{F}_p(\rho)$, we can compute the Hasse invariant, $H(\bar{J}_\rho)$, of $\bar{J}_\rho$. By the definition (see Hasse [1] or Manin [9]), $H(\bar{J}_\rho)$ is determined by the action of the Cartier operator ζ on $H^0(\bar{J}_\rho, \Omega^1)$ (which is closed under ζ) :

$$\zeta : H^0(\bar{J}_\rho, \Omega^1) \longrightarrow H^0(\bar{J}_\rho, \Omega^1)$$

$$\zeta(\bar{\omega}_\rho) = H(\bar{J}_\rho)^{(1/p)} \bar{\omega}_\rho$$

where $H(\bar{J}_\rho)^{(1/p)}$ denotes the p-th root of $H(\bar{J}_\rho)$ in $\mathbb{F}_p(\rho, \rho^{1/p^\infty})$. From the definition, it follows that $H(\bar{J}_\rho)$ is precisely given by the coefficient of $X^{p-1}\,dX$ in the expansion (I.1.3), reduced modulo pR. Therefore, we have

$$H(\bar{J}_\rho) \equiv P_{(p-1)/2}(\rho) \quad (\text{mod } pR). \tag{I.5.1}$$

Combining (I.5.1) with Proposition (I.4.4), we obtain the following result.

<u>(I.5.2) Corollary.</u>

$$[p](X) \equiv H(\bar{J}_\rho)X^p \quad (\underline{\text{mod}}\ pR,\ \underline{\text{mod}}\ \deg p+1),$$

<u>and accordingly,</u> $H(\bar{J}_\rho) \neq 0$ <u>in</u> $\mathbb{F}_p(\rho)$.

<u>(I.5.3) Proposition.</u> <u>Let</u> $F_{\bar{J}_\rho}$ <u>denote the formal group law of</u> J_ρ <u>over</u> $\mathbb{F}_p(\rho)$. <u>Then the weak isomorphism class of</u> $F_{\bar{J}_\rho}$ <u>over the algebraic closure</u> $\overline{\mathbb{F}_p(\rho)}$ <u>of</u> $\mathbb{F}_p(\rho)$ <u>is represented by the multiplicative formal group law</u> $\hat{\mathbb{G}}_m^{\pm}$.

<u>Proof</u>. Since $H(\bar{J}_\rho) \neq 0$, by Corollary (I.5.2), $F_{\bar{J}_\rho}$ has height $h = 1$. We know that over $\overline{\mathbb{F}_p(\rho)}$, height classifies weak isomorphism classes of formal group laws (Honda [3]). Obviously, $\hat{\mathbb{G}}_m^{\pm}$ is of height $h = 1$. Therefore, $F_{\bar{J}_\rho}$ is weakly isomorphic over $\overline{\mathbb{F}_p(\rho)}$ to $\hat{\mathbb{G}}_m^{\pm}$.

<u>I.6. Formal expansions of J_ρ and ω_ρ near zero: another parametrization.</u>

Let $t = X/Y$ and $w = X$, so that $X = w$ and $Y = w/t$. Then the equation (*) for the Jacobi quartic J_ρ in the affine (t,w)-plane is given by

$$w^2 = t^2(1 - 2\rho w^2 + w^4). \tag{I.6.1}$$

The point $(0,1)$ which is taken as the zero for the group law on J_ρ is transformed to the point $(0,0)$ in the (t,w)-plane, and $t = X/Y$ is a local parameter at this point. The recursive substitution of w^2 in the right hand side of (I.6.1) yields the formal expansion of J_ρ near zero, which starts off

$$w^2 = t^2 - 2\rho t^4 + (1+4\rho^2)t^6 - (6\rho + 8\rho^3)t^8$$
$$+ 2(1+4\rho+8\rho^2+8\rho^4)t^{10} + \cdots .$$

In fact, we have

$$w^2 = t^2 \sum_{n=0}^{\infty} A_{2n}(\rho)\, t^{2n}$$

where $A_{2n}(\rho)$ is a polynomial in 2ρ of degree n over $\mathbb{Z}$ for all $n \geq 0$, so $A_{2n} \in \mathbb{Z}[2\rho] \subset R$.

A formal expansion for the differential of the first kind ω_ρ of J_ρ near zero is obtained in the following manner. By differentiating (I.6.1), we get

$$dw = \frac{t(1-2\rho w^2 + w^4)}{w(1+2\rho t^2 - 2t^2w^2)}dt = \frac{w}{t}(1+2\rho t^2 - 2t^2w^2)^{-1}\ dt.$$

Then the formal power series expansion in t near zero of ω_ρ is

$$\omega_\rho = \frac{dX}{Y} = \frac{dw}{w/t} = \frac{dt}{1+2\rho t^2 - 2t^2w^2} = \sum_{n=0}^{\infty} (-2\rho+2w^2)^n\ t^{2n}\ dt$$

$$= dt\,\{ 1 - 2\rho t^2 + 2(1+2\rho^2)t^4 - 4(3\rho+2\rho^3)t^6$$
$$+ 2(3+24\rho^2+8\rho^4)t^8 + \cdots \}.$$

<u>(I.6.2) Proposition.</u>

$$\omega_\rho = \sum_{n=0}^{\infty} a_{2n}(\rho)\ t^{2n}\ dt$$

<u>where</u> $a_{2n}(\rho)$ <u>is a polynomial in</u> 2ρ <u>of degree</u> n <u>over</u> $\mathbb{Z}$ <u>for all</u> $n \geq 0$. <u>In particular,</u> $a_{2n}(\rho) \in \mathbb{Z}[2\rho] \subset R$.

In this parameterization, the Addition Theorem (I.3.1) on J_ρ is given as follows.

<u>(I.6.3) Proposition.</u> <u>With the notations of Theorem (I.3.1) in force, put</u>

$$t_1 = \frac{x(u)}{y(u)}, \quad w_1 = w(u) \quad ; \quad t_2 = \frac{x(v)}{y(v)}, \quad w_2 = x(v).$$

<u>Then</u>

$$x(u+v) = \frac{w_1 w_2}{t_1 t_2} \frac{t_1 + t_2}{1 - w_1^2 w_2^2}.$$

<u>In particular,</u>

$$x(2u) = \frac{w_1^2}{t_1^2} \frac{2t}{1 - w_1^4}.$$

In this parameterization, the formal group law of J_ρ takes the following form. Verification is left to the reader as an exercise.

<u>(I.6.4) Proposition.</u> <u>The formal group law of</u> J_ρ <u>is given by a formal power series</u> G_{J_ρ} <u>in two variables</u> T_1 <u>and</u> T_2 <u>over</u> $\mathbb{Z}[2\rho] \subset R$, <u>and it starts off</u>

$$\begin{aligned} G_{J_\rho}(T_1, T_2) &= T_1 + T_2 + 2\rho(T_1^2 T_2 + T_1 T_2^2) \\ &+ 2(2\rho^2 - 1)(T_1^4 T_2 + T_1 T_2^4) \\ &+ 4(2\rho^2 - 1)(T_1^2 T_2^3 + T_1^3 T_2^2) + \cdots \quad \in \mathbb{Z}[2\rho] \subset R. \end{aligned}$$

(G_{J_ρ} may be considered as the "dual" formal group law of F_{J_ρ} of Proposition (I.3.2).)

The Hasse invariant $H(\overline{J}_\rho)$ has the following representation in this parametrization, which follows directly from the definition.

<u>(I.6.5) Proposition.</u>

$$H(\overline{J}_\rho) \equiv a_{p-1}(\rho) \pmod{pR}.$$

(I.6.6) Proposition.

$$P_{(p-1)/2}(\rho) \equiv a_{p-1}(\rho) \pmod{pR}.$$

Proof. The Hasse invariant $H(\bar{J}_\rho)$ of $\bar{J}_\rho$ is independent of the choice of local parameters at zero. Equivalently, the Cartier operator $\mathcal{C}$ has a unique representation on $H^0(\bar{J}_\rho, \Omega^1)$ independently of the choice of local parameters at zero.

II. The Schur congruence and formal groups associated to Legendre polynomials

II.1. The Schur congruence.

Let $P_n(\rho)$ be the n-th Legendre polynomial. As we can see from (I.1.4), the r-th coefficient of $P_n(\rho)$ is a rational number $(-1)^r \binom{n}{r}\binom{2n-2r}{n-2r}$ with the denominator 2^n. Moreover, for an odd prime p and $i \geq 1$,

$$\left[\frac{2n-2r}{p^i}\right] \geq \left[\frac{n-r}{p^i}\right] + \left[\frac{n-r}{p^i}\right] \geq \left[\frac{n-r}{p^i}\right] + \left[\frac{n-2r}{p^i}\right] + \left[\frac{r}{p^i}\right]$$

where $[m]$ denotes the greatest integer function of $m \in \mathbb{Q}$. This means that $P_n(\rho) \in \mathbb{Z}[1/2][\rho] \subset R$. Note that $P_{(p-1)/2}(\rho)$ is a unit in R. We write $\bar{P}_n(\rho)$ for $P_n(\rho) \pmod{pR}$.

It is classically known that $P_n(\rho)$ satisfies the congruence due to Schur.

(II.1.1) Theorem (The Schur congruence for $P_n(\rho)$). If $n = a_0 + a_1 p + a_2 p^2 + \ldots + a_\mu p^\mu \in \mathbb{N}$ where $a_i < p$ for every i, then $P_n(\rho)$ satisfies the congruence

$$P_n(\rho) \equiv P_{a_0}(\rho)\, P_{a_1}(\rho^p)\, P_{a_2}(\rho^{p^2}) \cdots P_{a_\mu}(\rho^{p^\mu}) \pmod{pR}.$$

(II.1.2) Remarks. (a) The Schur congruence was first published in Ille [7] without proof ; the first published proof appeared in Wahab [10]. Wahab's proof rests on the following congruences

$$P_{mp+r}(\rho) \equiv P_r(\rho)\, P_{mp}(\rho) \pmod{pR} \quad \text{for} \quad r < p \quad \text{(Holt [2])},$$

and

$$P_{mp}(\rho) \equiv P_m(\rho^p) \pmod{pR} \quad \text{(Ille [7])}.$$

(b) There is a p-adic proof of the Schur congruence given by Landweber [8] (in this volume), which goes as follows. Observe that in $\mathbb{F}_p(\rho)[[X]]$ one has for any polynomial $f(X)$,

$$\frac{1}{\sqrt{1+f(X)}} = \lim_{\ell\to\infty} \{(1+f(X))^{(p-1)/2}\}^{1+p+p^2+\cdots+p^{\ell-1}}.$$

Applying this formula to $f(X) = -2\rho X^2 + X^4$, we obtain

$$\frac{1}{\sqrt{1-2\rho X^2+X^4}} = \sum_{n=0}^{\infty} P_n(\rho)\, X^{2n} = \Big(\sum_{n=0}^{p-1} P_n(\rho)\, X^{2n}\Big)^{1+p+p^2+\cdots}$$

in $\mathbb{F}_p(\rho)[[X]]$. Comparing the coefficients of X^{2n} of both sides, we get the Schur congruence (II.1.1).

(c) A special case of the Schur congruence for the Legendre polynomial $P_{(p^\mu-1)/2}(\rho)$ plays important roles in the subsequent discussions. Noting that $(p^\mu-1)/2 = (1+p+p^2+\cdots+p^{\mu-1})(p-1)/2$, the Schur congruence takes the form

$$\begin{aligned}\overline{P}_{(p^\mu-1)/2}(\rho) &= \overline{P}_{(p-1)/2}(\rho)\,\overline{P}_{(p-1)/2}(\rho^p)\cdots\overline{P}_{(p-1)/2}(\rho^{p^{\mu-1}})\\ &= \prod_{i=0}^{\mu-1}\overline{P}_{(p-1)/2}(\rho^{p^i})\end{aligned}$$

in $\mathbb{F}_p(\rho)$.

A geometrical (intrinsic) interpretation of the Schur congruence for $P_{(p^\mu-1)/2}(\rho)$ is described in the following Proposition.

(II.1.3) Proposition. Let

$$\mathcal{C}^{\mu} : H^0(\bar{J}_\rho, \Omega^1) \longrightarrow H^0(\bar{J}_\rho, \Omega^1)$$

denote the μ times iterated Cartier operator on $H^0(\bar{J}_\rho, \Omega^1)$. Then the following statements are equivalent.

(i) $\mathcal{C}^{\mu}$ has a unique representation with respect to a basis $\bar{\omega}_\rho$ ((I.1.3)) of $H^0(\bar{J}_\rho, \Omega^1)$.

(ii) The Schur congruence holds for $P_{(p^\mu-1)/2}(\rho)$, that is,

$$P_{(p^\mu-1)/2}(\rho) \equiv \prod_{i=0}^{\mu-1} P_{(p-1)/2}(\rho^{p^i}) \quad (\text{mod } pR).$$

Proof. $\mathcal{C}^{\mu}$ is represented on the one hand by

$$\mathcal{C}^{\mu}(\bar{\omega}_\rho) = \bar{P}_{(p^\mu-1)/2}(\rho)^{(1/p^\mu)}\,\bar{\omega}_\rho$$

and on the other hand by

$$\mathcal{C}^{\mu}(\bar{\omega}_\rho) = \bar{P}_{(p-1)/2}(\rho)^{(1/p)}\,\mathcal{C}^{\mu-1}(\bar{\omega}_\rho) = \dots$$

$$= \left(\prod_{i=0}^{\mu-1} \bar{P}_{(p-1)/2}(\rho^{p^i})\right)^{(1/p^\mu)} \bar{\omega}_\rho .$$

Now it is immediate to see that $\mathcal{C}^{\mu}$ has a unique representation with respect to a basis $\bar{\omega}_\rho$ of $H^0(\bar{J}_\rho, \Omega^1)$, if and only if the Schur congruence holds true for $P_{(p^\mu-1)/2}(\rho)$.

II.2. Formal groups associated to Legendre polynomials.

We shall study formal groups associated to the Jacobi quartic J_ρ in the framework of Honda's theory of formal groups (group laws). Observe first that $K = \mathbb{Q}(\rho)$ satisfies the condition (F_1) of Honda [3].

(F_1) [The valuation ν_p of K is unramified, and there exists an endomorphism σ of K such that $\alpha^\sigma \equiv \alpha^p$ (mod pR) for any $\alpha \in R$.]

Furthermore, for any $f(\rho) \in \mathbb{Z}[1/2][\rho] \subset R$, σ acts as

$$f^{\sigma}(\rho) \equiv f(\rho^{p}) \pmod{pR}.$$

In this context, the Schur congruence for $P_{(p^{\mu}-1)/2}(\rho)$ takes the following form. First we introduce the quantities $V_1, V_2, \ldots$ in R. Put

$$V_1 = P_{(p-1)/2}(\rho).$$

To define V_2, we make use of the Schur congruence for $P_{(p^2-1)/2}(\rho)$:

$$P_{(p^2-1)/2}(\rho) \equiv P_{(p-1)/2}(\rho)\, P^{\sigma}_{(p-1)/2}(\rho)$$

$$\equiv V_1^{1+\sigma} \pmod{pR}.$$

Now put

$$P_{(p^2-1)/2}(\rho) = V_1^{1+\sigma} + pV_2$$

To define V_3, we use the Schur congruence for $P_{(p^3-1)/2}(\rho)$. We have, on the one hand

$$P_{(p^3-1)/2}(\rho) \equiv P_{(p^2-1)/2}(\rho)\, P^{\sigma^2}_{(p-1)/2}(\rho)$$

$$\equiv V_1^{\sigma^2}\, P_{(p^2-1)/2}(\rho) \pmod{pR},$$

and on the other hand

$$P_{(p^3-1)/2}(\rho) \equiv P_{(p-1)/2}(\rho)\, P^{\sigma}_{(p^2-1)/2}(\rho) \pmod{pR}.$$

Comparing these two expressions, we define V_3 by putting

$$P_{(p^3-1)/2}(\rho) = V_1^{\sigma^2}\, P_{(p^2-1)/2}(\rho) + p\{V_1^{\sigma^2}\, P_{(p^2-1)/2}(\rho)$$

$$- P_{(p-1)/2}(\rho)\, P^{\sigma}_{(p^2-1)/2}(\rho)\} + p^2 V_3$$

$$= V_1^{\sigma^2} P_{(p^2-1)/2}(\rho) + pV_2^{\sigma} P_{(p-1)/2}(\rho) + p^2 V_3.$$

Inductively, V_μ can be defined. Now the Schur congruence for $P_{(p^\mu-1)/2}(\rho)$ takes the following form.

<u>(II.2.1) Proposition.</u> <u>With</u> $V_1, V_2, \ldots, V_\mu \in R$ <u>as above,</u>

$$P_{(p^\mu-1)/2}(\rho) = \sum_{\nu=0}^{\mu-1} p^{\mu-1-\nu} V_{\mu-\nu}^{\sigma^\nu} P_{(p^\nu-1)/2}(\rho).$$

Let $K_\sigma[[T]]$ (resp. $R_\sigma[[T]]$) denote the non-commutative power series ring over K (resp. R) with indeterminate T and the multiplication rule

$$T\alpha = \alpha^\sigma T \quad \text{for any} \quad \alpha \in K.$$

For any $f(x) \in K[[x]]$ and $u(T) = \sum_{\mu=0}^{\infty} C_\mu T^\mu \in K_\sigma[[T]]$, we define an element $u*f$ of $K[[x]]$ by

$$(u*f)(x) = \sum_{\mu=0}^{\infty} C_\mu f^{\sigma^\mu}(x^{p^\mu}).$$

An element $u(T) \in R_\sigma[[T]]$ is said to be <u>special</u> if $u(T) \equiv p$ (mod deg 1). An element $f(x) \in K[[x]]$ is said to be <u>of type</u> $(C;u)$ with C a unit in R and u a special element, if

$$f(x) \equiv C\,x \quad (\text{mod} \quad \deg \quad 2),$$

and

$$(u*f)(x) \equiv 0 \ (\text{mod} \quad pR).$$

The formal group law F_{J_ρ} of J_ρ (Proposition (I.3.2)) can be restated in the framework of Honda's theory as follows.

<u>(II.2.2) Proposition.</u> <u>Let</u>

$$f_\rho(X) = \sum_{n=0}^{\infty} \frac{P_n(\rho)}{2n+1} X^{2n+1} \quad \in K[[X]]$$

<u>and form</u>

$$F(X_1,X_2) = f_\rho^{-1}(f_\rho(X_1) + f_\rho(X_2)).$$

Then F is a formal group law defined over R, which is the formal group law F_{J_ρ} of J_ρ. In other words, $f_\rho(X)$ is the logarithm of F_{J_ρ}.

A p-typical version of Proposition (II.2.2) is given as follows. Here p is a fixed odd prime.

(II.2.3) Proposition. Let

$$f_\rho(X) = \sum_{\mu=0}^{\infty} P_{(p^\mu-1)/2}(\rho)\, p^{-\mu} X^{p^\mu} \in K[[X]].$$

Then there eixsts a special element

$$u(T) = p + \sum_{\nu=1}^{\infty} C_\nu T^\nu \in R_\sigma[[T]]$$

such that f_ρ is of type $(1\,;u)$, i.e., $(u*f_\rho)(X) \equiv 0 \pmod{pR}$, and that

$$f_\rho^{-1}(f_\rho(X_1) + f_\rho(X_2))$$

is the formal group law F_{J_ρ} of J_ρ.

(II.2.4) Proposition. With $f_\rho(X)$ and $u(T)$ as in Proposition (II.2.3), put

$$(u*f_\rho)(X) = \sum_{\mu=0}^{\infty} pU_\mu X^{p^\mu}$$

with $U_0 = 1$ and $U_\mu \in R$ for every $\mu \geq 1$. Then

$$pU_\mu = \frac{1}{p^{\mu-1}}\left\{ P_{(p^\mu-1)/2}(\rho) + \sum_{\nu=1}^{\mu} p^{\nu-1} C_\nu P^{\sigma^\nu}_{(p^{\mu-\nu}-1)/2}(\rho) \right\} \in R$$

for every $\mu \geq 1$. Equivalently,

$$P_{(p^\mu-1)/2}(\rho) + \sum_{\nu=1}^{\mu} p^{\nu-1} C_\nu P^{\sigma^\nu}_{(p^{\mu-\nu}-1)/2}(\rho) \equiv 0 \pmod{p^\mu R}$$

for every $\mu \geq 1$.

Furthermore, combining with the Schur congruence (Proposition (II.2.1)), we have

$$V_{\mu-\nu}^{\sigma^{\nu}} P_{(p^{\nu}-1)/2}(\rho) + C_{\mu-\nu} P^{\sigma^{\mu-\nu}}_{(p^{\nu}-1)/2}(\rho) \equiv 0 \pmod{p^{\nu+1}R}$$

for every ν, $0 \leq \nu \leq \mu-1$.

Now we obtain a main result of this section, which gives an intrinsic interpretation of the Schur congruence for $P_{(p^{\mu}-1)/2}(\rho)$.

(II.2.5) Theorem. With the notations and the hypothesis of Propositions (II.2.2), (II.2.3) and (II.2.4) in force, the following statements are equivalent.

(i) The integrality of the formal group law F_{J_ρ} of J_ρ at ν_p, that is, $F_{J_\rho}(X_1,X_2) \in R[[X_1,X_2]]$.

(ii) The validity of the Schur congruence for $P_{(p^{\mu}-1)/2}(\rho)$:

$$P_{(p^{\mu}-1)/2}(\rho) \equiv \prod_{i=0}^{\mu-1} P_{(p-1)/2}(\rho^{p^i}) \pmod{pR}$$

for every $\mu \geq 0$.

Proof. (i) $\Rightarrow$ (ii) This follows from Propositions (II.2.2), (II.2.3) and (II.2.4).

(ii) $\Rightarrow$ (i) The Schur congruence for $P_{(p^{\mu}-1)/2}(\rho)$ guarantees the existence of a special element $u(T) \in R_\sigma[[T]]$ such that the logarithm $f_\rho(X) \in K[[X]]$ of the formal group law $F_{J_\rho}(X_1,X_2)$ is of type $(1\,;u)$. Then by Theorem 2 of Honda [3], we can conclude that $F_{J_\rho}(X_1,X_2)$ is defined over R.

II.3. The structure theorem for the formal group law F_{J_ρ} of J_ρ.

We can determine the structure of the formal group law F_{J_ρ} of J_ρ up to isomorphism (resp. weak isomorphism) over R and over certain extensions of R.

<u>(II.3.1) Proposition. The isomorphism class over R of the formal group law F_{J_ρ} of J_ρ is uniquely represented by the special element</u>

$$u(T) = p + \sum_{\nu=1}^{\infty} C_\nu T^\nu$$

$$= p + (pU_1 - P_{(p-1)/2}(\rho))T$$

$$+ (pU_2 - U_1 P^{\sigma}_{(p-1)/2}(\rho) - V_2)T^2 + \cdots \in R_\sigma[[T]]$$

<u>where V_μ and U_μ are elements of R defined as in Proposition (II.2.1) and Proposition (II.2.4), respectively.</u>

<u>Proof.</u> This is just a special case of Theorem 4 in Honda [3].

Let $\hat{R}$ denote the completion of R with respect to v_p. Then over $\hat{R}$, we have a finer structure theorem for the formal group law F_{J_ρ} of J_ρ.

<u>(II.3.2) Theorem. The isomorphism class over $\hat{R}$ of the formal group law F_{J_ρ} of J_ρ is uniquely represented by the special element</u>

$$\hat{u}(T) = p - P_{(p-1)/2}(\rho)T.$$

<u>In other words, the Hasse invariant $H(\bar{J}_\rho) \equiv P_{(p-1)/2}(\rho)$ of $\bar{J}_\rho$ is a "moduli" of the isomorphism class of F_{J_ρ} over $\hat{R}$.</u>

<u>Proof.</u> With $u(T)$ as in Proposition (II.3.1), we can find a unit $t(T) = 1 + \sum_{\nu=1}^{\infty} A_\nu T^\nu$ in $R_\sigma[[T]]$ such that

$$t(T)u(T) = p + \sum_{\nu=1}^{\infty} (pA_\nu + \sum_{\tau<\nu} A_\tau C_{\nu-\tau})T^\nu$$

$$= p - P_{(p-1)/2}(\rho)\, T.$$

In fact, we can choose $A_1, A_2, \ldots,$ sucessively and uniquely so that the above identity holds true. Therefore, the isomorphism class of F_{J_ρ} over $\hat{R}$ is uniquely represented by the special element $\hat{u}(T)$ $= p - P_{(p-1)/2}(\rho)\, T$. Thus, we may view the Hasse invariant $H(\bar{J}_\rho)$ $\equiv P_{(p-1)/2}(\rho)$ as a "moduli" of the isomorphism class of F_{J_ρ} over $\hat{R}$.

Let $\bar{R}$ denote the integral closure of R in the algebraic closure $\bar{K}$ of K, with the maximal ideal $p\bar{R}$. Observe that $\bar{K}$ satisfies the condition (F_1) described in II.2. Passing to $\bar{R}$, we obtain the following structure theorem for the formal group law F_{J_ρ} of J_ρ.

(II.3.3) Theorem. <u>The weak isomorphism class over $\bar{R}$ of the formal group law F_{J_ρ} of J_ρ is represented by the multiplicative formal group law $\hat{\mathbb{G}}_m^{\pm}$.</u>

Proof. By Proposition (I.5.3), one knows that the set $\mathrm{Iso}_{\overline{\mathbf{F}_p(\rho)}}(F_{\bar{J}_\rho}, \hat{\mathbb{G}}_m^{\pm}) \neq \emptyset$. Then by Honda [3] (4.3), every element $\phi \in \mathrm{Iso}_{\overline{\mathbf{F}_p(\rho)}}(F_{\bar{J}_\rho}, \hat{\mathbb{G}}_m^{\pm})$ is of the form $\phi = g^{-1} \circ (Cf) \pmod{p\bar{R}}$ where f and g are logarithms of F_{J_ρ} and $\hat{\mathbb{G}}_m^{\pm}$, respectively, and C is a unit in $\bar{R}$. This, in particular, implies that $\mathrm{Iso}_{\bar{R}}(F_{J_\rho}, \hat{\mathbb{G}}_m^{\pm}) \neq \emptyset$, and accordingly, F_{J_ρ} is weakly isomorphic to $\hat{\mathbb{G}}_m^{\pm}$ over $\bar{R}$.

II.4. The Atkin and Swinnerton-Dyer type congruence for the Jacobi quartic J_ρ modulo pR.

Let $\bar{\omega}_\rho$ denote the differential of the first kind of the reduced Jacobi quartic $\bar{J}_\rho$. $\bar{\omega}_\rho$ has the formal power series expansions as in (I.1.3) and (I.6.2) with coefficients reduced modulo pR. These expansions are p-typically read as follows (p an odd prime):

$$\text{(II.4.1)} \qquad \bar{\omega}_\rho = \sum_{\mu=0}^{\infty} \bar{P}_{(p^\mu-1)/2}(\rho)\, X^{p^\mu-1}\, dX,$$

$$\text{(II.4.2)} \qquad \bar{\omega}_\rho = \sum_{\mu=0}^{\infty} \bar{a}_{p^\mu-1}(\rho)\, t^{p^\mu-1}\, dt$$

where $\bar{P}_{(p^\mu-1)/2}(\rho) = P_{(p^\mu-1)/2}(\rho) \pmod{pR}$ and $\bar{a}_{p^\mu-1}(\rho) = a_{p^\mu-1}(\rho) \pmod{pR}$.

Comparing these two expressions for $\bar{\omega}_\rho$, we obtain the following congruence.

<u>(II.4.3) Proposition.</u> <u>For every</u> $\mu \geq 0$,

$$P_{(p^\mu-1)/2}(\rho) \equiv a_{p^\mu-1}(\rho) \pmod{pR}.$$

<u>Proof</u>. Apply the Cartier operator $\mathcal{C}$ to $\bar{\omega}_\rho$ in both expressions (II.4.1) and (II.4.2), considering $\bar{J}_\rho$ and $\bar{\omega}_\rho$ over the perfect closure $\mathbb{F}_p(\rho, \rho^{1/p^\infty})$ of $\mathbb{F}_p(\rho)$. We have from (II.4.1),

$$\mathcal{C}(\bar{\omega}_\rho) = \sum_{\mu=0}^{\infty} \bar{P}_{(p^{\mu+1}-1)/2}(\rho)^{(1/p)} X^{p^\mu} \frac{dX}{X}$$

$$= \bar{P}_{(p-1)/2}(\rho)^{(1/p)} \bar{\omega}_\rho,$$

and from (II.4.2),

$$\mathcal{C}(\bar{\omega}_\rho) = \sum_{\mu=0}^{\infty} \bar{a}_{p^{\mu+1}-1}(\rho)^{(1/p)} t^{p^\mu} \frac{dt}{t}$$

$$= \bar{a}_{p-1}(\rho)^{(1/p)} \bar{\omega}_\rho$$

by the definition of the Hasse invariant of $\bar{J}_\rho$. (Cf. (I.5).)

Now we let the Cartier operator $\mathcal{C}$ act μ times repeatedly on $\bar{\omega}_\rho$, noting that $H^0(\bar{J}_\rho, \Omega^1)$ is closed under the iterated action of $\mathcal{C}$. We have, on the one hand

$$\mathcal{C}^\mu(\bar{\omega}_\rho) = P_{(p^\mu-1)/2}(\rho)^{(1/p^\mu)} \bar{\omega}_\rho.$$

We have, on the other hand

$$\mathcal{C}^\mu(\bar{\omega}_\rho) = \bar{a}_{p-1}(\rho)^{(1/p)} \mathcal{C}^{\mu-1}(\bar{\omega}_\rho) = \cdots$$

$$= \left(\bar{a}_{p-1}(\rho)\, \bar{a}_{p-1}(\rho^p) \cdots \bar{a}_{p-1}(\rho^{p^{\mu-1}})\right)^{(1/p^\mu)} \bar{\omega}_\rho.$$

In the above expressions, $\alpha^{(1/p^m)}$ denotes the p^m-th root of an element $\alpha \in \mathbb{F}_p(\rho)$ in the perfect closure $\mathbb{F}_p(\rho, \rho^{1/p^\infty})$ of $\mathbb{F}_p(\rho)$.

Then it follows from Proposition (I.6.6) that ζ^{μ} has a unique representation independently of the choice of local parameters at zero. In return, this implies that

$$\overline{P}_{(p^{\mu}-1)/2}(\rho)^{(1/p^{\mu})} = \overline{a}_{p^{\mu}-1}(\rho)^{(1/p^{\mu})}$$

and this yields the required congruence.

(II.4.4) Theorem (The Atkin and Swinnerton-Dyer type congruence for J_{ρ} modulo pR). Let

$$\omega_{\rho} = \sum_{n=0}^{\infty} a_{2n}(\rho)\, t^{2n}\, dt$$

be the formal power series expansion of the first kind differential 1-form ω_{ρ} on J_{ρ} with respect to a local parameter t at zero. Let p be an odd prime. Then $a_{p^{\mu}-1}(\rho)$ satisfies the congruence

$$a_{p^{\mu}-1}(\rho) \equiv a_{p-1}(\rho)\, a_{p-1}(\rho^{p}) \ldots a_{p-1}(\rho^{p^{\mu-1}}) \pmod{pR}$$

for every $\mu \geq 0$.

Proof. This follows immediately from Theorem (II.1.1), Proposition (II.4.3) and Proposition (I.6.6).

We now formulate the congruence for $a_{p^{\mu}-1}(\rho)$ in a form corresponding to the Schur congruence for $P_{(p^{\mu}-1)/2}(\rho)$ in Proposition (II.2.1).

(II.4.5) Proposition.

$$a_{p^{\mu}-1}(\rho) = \sum_{\nu=0}^{\mu-1} p^{\mu-1-\nu}\, W_{\mu-\nu}^{\sigma^{\nu}}\, a_{p^{\nu}-1}(\rho)$$

where $W_1, W_2, \ldots$ are elements of R defined analogously as for $V_1, V_2, \ldots,$ using $a_{p^{\nu}-1}(\rho)$ in place of $P_{(p^{\nu}-1)/2}(\rho)$ for every $\nu \geq 0$.

II.5. The formal group law of J_ρ : another parametrization.

Let

$$g_\rho(t) = \sum_{n=0}^{\infty} \frac{a_{2n}(\rho)}{2n+1} t^{2n+1} \in K[[t]],$$

or p-typically, for an odd prime p, let

$$g_\rho(t) = \sum_{\mu=0}^{\infty} a_{p^\mu - 1}(\rho)\, p^{-\mu}\, t^{p^\mu} \in K[[t]].$$

Form

$$G(T_1, T_2) = g_\rho^{-1}(g_\rho(T_1) + g_\rho(T_2)).$$

Then G is a formal group defined over R, which is the formal group law G_{J_ρ} of J_ρ of Proposition (I.6.4). There exists a special element

$$v(T) = p + \sum_{\nu=1}^{\infty} D_\nu T^\nu \in R_\sigma[[T]]$$

such that

$$(v*g_\rho)(t) \equiv 0 \pmod{pR}.$$

Put

$$(v*g_\rho)(t) = \sum_{\mu=0}^{\infty} pS_\mu t^{p^\mu} \in pR[[t]].$$

Then $v(T)$ starts off

(II.5.1) $$v(T) = p + (pS_1 - a_{p-1}(\rho))T + (pS_2 - S_1 a_{p-1}^\sigma(\rho) - W_2)T^2 + \cdots \in R_\sigma[[T]],$$

and it represents the isomorphism class of G over R. Passing to the completion $\hat{R}$ of R, the isomorphism class of G is represented by the special element

$$\hat{v}(T) = p - a_{p-1}(\rho)\, T,$$

that is, a "moduli" of G is the Hasse invariant $H(\bar{J}_\rho) \equiv a_{p-1}(\rho)$ of the reduced Jacobi quartic $\bar{J}_\rho$.

(II.5.2) Proposition. The formal group laws G and hence G_{J_ρ} are isomorphic over R to the formal group law F of Proposition (II.2.2) and hence to F_{J_ρ} of Proposition (I.3.2).

Consequently, the structure of the isomorphism class of the formal group law of J_ρ is independent of the choice of local parameters at zero of J_ρ.

Proof. By Honda [3], Theorem 3, F_{J_ρ} and G_{J_ρ} are isomorphic over R, if and only if there exists a unit $t(T) = 1 + \sum_{\nu=1}^{\infty} B_\nu T^\nu \in R_\sigma[[T]]$ such that $t(T)u(T) = v(T)$ where $u(T)$ (resp. $v(T)$) is the special element associated to F_{J_ρ} (resp. G_{J_ρ}) explicitly given in Proposition (II.3.1) (resp. (II.5.1)). Now we claim that the equation (with $u(T) = p + \sum_{\nu=1}^{\infty} C_\nu T^\nu$)

$$t(T)u(T) = p + \sum_{\nu=1}^{\infty} (pB_\nu + \sum_{\tau<\nu} B_\tau C_{\nu-\tau}^{\sigma^\tau})T^\nu = v(T)$$

can be solved uniquely for $B_1, B_2, \ldots$ in R. Indeed the existence of such solutions follows from Proposition (II.4.3).

III. The Honda congruences and their application.

III.1. The Honda congruences (Reference : Honda [4]).

Honda has obtained two congruences satisfied by the Legendre polynomial $P_n(\rho)$ modulo higher powers of p. Unfortunately, an error slipped into his argument, namely, he regarded "$\mathbb{Z}_p[\rho]$ as a discrete valuation ring". This error, however, can be corrected easily considering $P_n(\rho)$ over the discrete valuation ring R. Under this situation, the Honda theory of formal groups [3] can be applied and the congruences of Honda are still valid.

(III.1.1) Theorem (The Honda congruences for $P_n(\rho)$). For any $m \geq 1$ and $\mu \geq 0$, we have

(a) $P_{mp^{\mu+1}}(\rho) \equiv P_{mp^{\mu}}(\rho^p) \pmod{p^{\mu+1}R}$,

(b) $P_{mp^{\mu+1}-1}(\rho) \equiv P_{mp^{\mu}-1}(\rho^p) \pmod{p^{\mu+1}R}$.

We now apply the same line of arguments as in (II.1.2)(b) to the function $(1-2\rho X+X^2)^{-1/2}$ and get

$$\frac{1}{\sqrt{1-2\rho X+X^2}} = \sum_{n=0}^{\infty} P_n(\rho)\, X^n.$$

Furthermore, modulo pR, we obtain

$$\text{(III.1.2)} \qquad \sum_{n=0}^{\infty} P_n(\rho)\, X^n \equiv \left(\sum_{n=0}^{p-1} P_n(\rho)\, X^n\right)^{1+p+p^2+\cdots}.$$

(III.1.3) Remarks. (a) Comparing the coefficients of $X^{mp^{\mu}}$ of both sides in (III.1.2), we obtain the congruence

$$P_{mp^{\mu}}(\rho) \equiv P_m(\rho^{p^{\mu}}) \pmod{pR}$$

for every $\mu \geq 0$. It is easy to see that this congruence can be deduced from the Honda congruence (a). In fact,

$$\begin{aligned} P_{mp^{\mu}}(\rho) &\equiv P_{mp^{\mu-1}}(\rho^p) \pmod{p^{\mu}R} \\ &\equiv P_{mp^{\mu-2}}(\rho^{p^2}) \pmod{p^{\mu-1}R} \\ &\cdots\cdots\cdots\cdots\cdots\cdots \\ &\equiv P_m(\rho^{p^{\mu}}) \pmod{pR}. \end{aligned}$$

(b) Comparing the coefficients of $X^{mp^{\mu}-1}$ of both sides of (III.1.2), we also get the Schur congruence for $P_{mp^{\mu}-1}(\rho)$. In fact,

$$P_{mp^{\mu}-1}(\rho) \equiv P_{p-1}(\rho)\, P_{mp^{\mu-1}-1}(\rho^p) \pmod{p^{\mu}R}$$

$$\equiv P_{p-1}(\rho)\, P_{p-1}(\rho^p)\, P_{mp^{\mu-2}-1}(\rho^{p^2}) \pmod{p^{\mu-1}R}$$

$$\cdots\cdots\cdots\cdots\cdots\cdots\cdots$$

$$\equiv P_{p-1}(\rho)\, P_{p-1}(\rho^p) \ldots P_{p-1}(\rho^{p^{\mu-1}})\, P_{m-1}(\rho^{p^\mu}) \pmod{pR}$$

for every $\mu \geq 0$.

(c) We look into the ratio $P_{mp^{\mu+1}-1}(\rho)/\,P_{mp^\mu-1}(\rho^p)$, which we denote by T. On the one hand, the Schur congruence gives

$$T \equiv P_{p-1}(\rho) \pmod{pR},$$

and on the other hand, the Honda congruence (b) gives

$$T \equiv 1 \pmod{p^{\mu+1}R}.$$

Noting that

$$P_{p-1}(\rho) \equiv 1 \pmod{pR},$$

we get

$$T = \frac{P_{mp^{\mu+1}-1}(\rho)}{P_{mp^\mu-1}(\rho^p)} \equiv (P_{p-1}(\rho))^{p^\mu} = 1 \pmod{p^{\mu+1}R}$$

for every $\mu \geq 0$.

III.2. The formal group law of the rational curve C_ρ.

The Honda congruences reflect certain geometrical properties of the rational curve

$$C_\rho \quad : \qquad Y^2 = 1 - 2\rho X + X^2$$

over $K = \mathbb{Q}(\rho)$. The function field, $K(C_\rho)$, of C_ρ over K is the subfield of the function field, $K(J_\rho)$, of the Jacobi quartic J_ρ over K, which consists of all even functions. If $K(J_\rho) = K(X,Y)$, then $K(C_\rho) = K(X^2,Y)$ where (X,Y) denotes the coordinate functions on J_ρ.

The formal group law of C_ρ is given as follows.

(III.2.1) Proposition. (Cf. Honda [4].) Let

$$h_\rho(X) = \sum_{n=1}^{\infty} \frac{P_n(\rho)}{n+1} X^{n+1} \in K[[X]].$$

Then $h_\rho(X)$ is of type $u(T) = p - T \in R_\sigma[[T]]$. Put

$$H(X_1, X_2) = h_\rho^{-1}(h_\rho(X_1) + h_\rho(X_2)).$$

Then H is a formal group defined over R, which is the formal group law of the rational curve C_ρ.

Furthermore, H is isomorphic over R to the multiplicative formal group law $\hat{\mathbb{G}}_m^{\pm}$.

Proof. The coefficient of $X^{mp^{\mu+1}}$ in $(u*h_\rho)(X)$ is given by

$$(mp^\mu)^{-1}\{P_{mp^{\mu+1}-1}(\rho) - P_{mp^\mu-1}(\rho^p)\} \equiv 0 \pmod{pR}$$

by the Honda congruence (III.1.1)(b). Hence $h_\rho(X)$ is of type $(1; p-T)$ and H is a formal group law defined over R, isomorphic over R to $\hat{\mathbb{G}}_m^{\pm}$ by Theorems 2 and 3 of Honda [3].

(III.2.2) Remark. The Jacobi quartic J_ρ is the double cover of the rational curve C_ρ over K. We compare the formal group laws of both curves.

	C_ρ	J_ρ
logarithm of formal group law	$\sum_{n=0}^{\infty} \frac{P_n(\rho)}{n+1} X^{n+1}$	$\sum_{n=0}^{\infty} \frac{P_n(\rho)}{2n+1} X^{2n+1}$
invariant differential	$\sum_{n=0}^{\infty} P_n(\rho) X^n \, dX$	$\sum_{n=0}^{\infty} P_n(\rho)\, X^{2n} \, dX$
height of formal group law	1	1
A special element over R	$p - T$	$p - P_{(p-1)/2}(\rho)T + \cdots$
isomorphism class of formal group law	$\hat{\mathbb{G}}_m^{\pm}$ over $\bar{R}$ over $\hat{R}$	$\hat{\mathbb{G}}_m^{\pm}$ height 1 formal group law parametrized by $P_{(p-1)/2}(\rho)$

Appendix : The formal group associated to the twisted Legendre polynomials

We shall construct a formal group associated to twisted Legendre polynomials. Its geometrical interpretation in connection with the Jacobi quartic J_ρ, will not be discussed here, but is left for the future investigation.

The notations of III.1 remain in force throughout this appendix. Theorems cited below are essentially due to Honda [4].

(A.1) Theorem. *Let*

$$f_\alpha(X) = \sum_{\mu=0}^{\infty} P^{\sigma^\mu-1}_{(p-1)/2}(\rho)\ p^{-\mu}\ X^{p^\mu} \quad \varepsilon\ K[[X]].$$

Then $f_\alpha(X)$ *is of type* $(1\ ;\ u_\alpha)$ *where*

$$u_\alpha(T) = p - P^{\sigma-1}_{(p-1)/2}(\rho)\ T \ \varepsilon\ R_\sigma[[T]]$$

is a special element. Form

$$F_\alpha(X_1, X_2) = f_\alpha^{-1}(f_\alpha(X_1) + f_\alpha(X_2)).$$

Then F_α *is a formal group law defined over* R. *Moreover,* F_α *is a generalized Lubin-Tate formal group.*

Proof. We have for every $\mu \geq 0$,

$$P^{\sigma^{\mu+1}-1}_{(p-1)/2}(\rho) = P^{\sigma-1}_{(p-1)/2}(\rho)\ P^{\sigma(\sigma^\mu-1)}_{(p-1)/2}(\rho).$$

Then it follows that

$$(u_\alpha * f_\alpha)(X) = pX + \sum_{\mu=0}^{\infty}\left(P^{\sigma^{\mu+1}-1}_{(p-1)/2}(\rho) - P^{\sigma-1}_{(p-1)/2}(\rho)\ P^{\sigma(\sigma^\mu-1)}_{(p-1)/2}(\rho)\right)$$

$$\times\ p^{-\mu}\ X^{p^{\mu+1}} = pX \equiv 0 \pmod{pR}.$$

Hence $f_\alpha(X)$ is of type $(1\ ;\ u_\alpha)$ and Honda [3], Theorem 2 again asserts that F_α is a formal group law with coefficients in R. To show the last assertion, we note that

$$pf_\alpha(X) = (u_\alpha * f_\alpha)(X) + P^{\sigma-1}_{(p-1)/2}(\rho)\ f^\sigma_\alpha(X^p)$$

$$= pX + P^{\sigma-1}_{(p-1)/2}(\rho)\ f^\sigma_\alpha(X^p).$$

Then the endomorphism $[p]_{F_\alpha}$ (multiplication by p on F_α) satisfies the congruence

$$[p]_{F_\alpha}(X) \equiv pX \pmod{\deg 2}$$

and

$$[p]_{F_\alpha}(X) = f_\alpha^{-1}(pf_\alpha(X))$$

$$\equiv f_\alpha^{-1}(P^{\sigma-1}_{(p-1)/2}(\rho)\ f^\sigma_\alpha(X^p)) \pmod{pR}$$

$$\equiv P^{\sigma-1}_{(p-1)/2}(\rho)\ X^p \pmod{pR}.$$

Indeed, the last congruence follows from the following observation. Note that for every $\mu \geq 1$,

$$\sigma^\mu - 1 = (\sigma - 1)(1 + \sigma + \sigma^2 + \dots + \sigma^{\mu-1}).$$

The coefficient of $p^{-\mu} X^{p^{\mu+1}}$ in $P^{\sigma-1}_{(p-1)/2}(\rho)\ f^\sigma_\alpha(X^p)$ is

$$P^{\sigma-1}_{(p-1)/2}(\rho)\ P^{\sigma(\sigma^\mu-1)}_{(p-1)/2}(\rho)$$

$$\equiv P^{\sigma-1}_{(p-1)/2}(\rho)\ P^{\sigma^\mu-1}_{(p-1)/2}(\rho^p) \pmod{pR}$$

....

$$\equiv P^{\sigma-1}_{(p-1)/2}(\rho)\ P^{\sigma-1}_{(p-1)/2}(\rho^p) \dots P^{\sigma-1}_{(p-1)/2}(\rho^{p^\mu}) \pmod{pR},$$

while the coefficient of $p^{-\mu} X^{p^{\mu+1}}$ in $f_\alpha(P^{\sigma-1}_{(p-1)/2}(\rho)\ X^p)$ is

$$P^{\sigma^\mu-1}_{(p-1)/2}(\rho)\ P^{(\sigma-1)p^\mu}_{(p-1)/2}(\rho)$$

$$\equiv P^{\sigma^\mu-1}_{(p-1)/2}(\rho)\ P^{\sigma-1}_{(p-1)/2}(\rho^{p^\mu}) \pmod{pR}$$

....

$$\equiv P^{\sigma-1}_{(p-1)/2}(\rho)\, P^{\sigma-1}_{(p-1)/2}(\rho^p) \ldots P^{\sigma-1}_{(p-1)/2}(\rho^{p^\mu}) \pmod{pR}.$$

Hence F_α is a formal group law of the generalized Lubin-Tate type for which $[p]_{F_\alpha}$ is an endomorphism.

(A.2) Theorem. *The formal group law F_α constructed in Theorem (A.1) is isomorphic over R to the multiplicative formal group law $\hat{\mathbb{G}}^{\pm}_m$.*

Proof. Let

$$g^{\pm}(X) = \pm \log(1 \pm X) = \pm \sum_{n=1}^{\infty} (-1)^{n-1} \frac{(\mp 1)^n X^n}{n}.$$

It is well known that $g^{\pm}(X)$ is of type $(1;\ v)$ where $v(T) = p - T \in R_\sigma[[T]]$ is a special element, and that it is the logarithm of $\hat{\mathbb{G}}^{\pm}_m$, i.e.,

$$g^{\pm} \circ (\hat{\mathbb{G}}^{\pm}_m) \circ (g^{\pm})^{-1} = \hat{\mathbb{G}}_a.$$

Now we have

$$\begin{aligned} P_{(p-1)/2}(\rho)\, u_\alpha(T) &= p\, P_{(p-1)/2}(\rho) - P^{\sigma}_{(p-1)/2}(\rho) T \\ &= (p - T)\, P_{(p-1)/2}(\rho) \\ &= v(T)\, P_{(p-1)/2}(\rho). \end{aligned}$$

Hence by Honda [3], Theorem 3, F_α is isomorphic over R to $\hat{\mathbb{G}}^{\pm}_m$. An explicit isomorphism from F_α into $\hat{\mathbb{G}}^{\pm}_m$ over R is given by

$$(g^{\pm})^{-1}(P_{(p-1)/2}(\rho)\, f_\alpha(X))$$

$$= 1 - \exp(P_{(p-1)/2}(\rho)\, f_\alpha(X)) \in R[[X]].$$

References

[1] Hasse, H., Existenz separabler zyklischer unverzweigter Erweiterrungskörper vom Primzahlgrade p über elliptischen Funktionenkorpern der Charakteristik p, J. Reine Angew. Math. 172 (1934), pp. 77-85.

[2] Holt, J.B., On the irreducibility of Legendre polynomials II, Proc. London Math. Soc. 2, Vol. 12 (1913), pp. 126-132.

[3] Honda, T., On the theory of commutative formal groups, J. Math. Soc. Japan 22 (1970), pp. 213-246.

[4] Honda, T., Two congruence properties of Legendre polynomials, Osaka J. Math. 13 (1976), pp. 131-133.

[5] Igusa, J., On the transformation theory of elliptic functions, Amer. J. Math. 81 (1959), pp. 436-452.

[6] Igusa, J., On the algebraic theory of elliptic modular functions, J. Math. Soc. Japan 20 (1968), pp. 96-106.

[7] Ille, H., Zur Irreduzibilität der Kugenfunktionen, Jahrbuch der Dissertationen der Universität Berlin 1924.

[8] Landweber, P., Supersingular elliptic curves and congruences for Legendre polynomials, in this volume.

[9] Manin, Ju.I., The Hasse-Witt matrix of an algebraic curve, Amer. Math. Soc. Transl. Ser. 45 (1965), pp. 245-264.

[10] Wahab, J.H., New cases of irreducibility for Legendre polynomials, Duke J. Math. 19 (1952), pp. 165-176.

[11] Whittaker, E.T., and Watson, G.N., A Course in Modern Analysis, Fourth edition, Cambridge University Press 1927.

Note on the Landweber-Stong elliptic genus

by Don Zagier

University of Maryland, College Park, MD 20742

Max-Planck-Institut für Mathematik, 5300 Bonn, FRG

In algebraic topology one studies genera, which are ring homomorphisms from the oriented bordism ring Ω_*^{SO} to a $\mathbb{Q}$-algebra R (commutative, with 1). To a genus $\varphi\colon \Omega_*^{SO} \to R$ one associates the following three power series with coefficients in R:

(i) $g(x)$, an odd power series with leading term x, the logarithm of the formal group law of φ (this means that the formal group, whose definition we do not repeat here, equals $g^{-1}(g(x)+g(y))$). It is given explicitly by $g(x) = \sum_{n=0}^{\infty} \varphi(\mathbb{CP}^{2n}) \frac{x^{2n+1}}{2n+1}$ and hence, since the classes of the $\mathbb{CP}^{2n}$ generate $\Omega_*^{SO}\otimes\mathbb{Q}$, determines φ completely.

(ii) $P(u)$, an even power series with leading term 1, the Hirzebruch characteristic power series of φ. This means that if $\mathcal{P}$ denotes the stable $H^*(\cdot;R)$-valued exponential characteristic class on oriented bundles characterized by $\mathcal{P}(\xi) = P(c_1(\xi))$ if ξ is a complex line bundle (regarded as a real 2-plane bundle), then the genus of an arbitrary oriented manifold M is obtained by evaluating $\mathcal{P}(TM)$ on the homology fundamental class of M.

(iii) $F(y)$, a power series with leading term 1, the KO-theory characteristic power series of φ. This means that if $\mathcal{F}$ denotes the stable $KO(\cdot)\otimes R$-valued exponential characteristic class on oriented bundles characterized by $\mathcal{F}(\xi) = F(\xi-[2])$ for ξ as above (this makes sense because $\xi-[2]$ is nilpotent in $KO(B_\xi)\otimes R$, as one sees by applying the complexified Chern character), then the genus of an arbitrary Spin manifold is obtained by evaluating $\mathcal{F}(TM)$ on a certain KO_*-fundamental class of M.

These three power series determine one another by the formulas

$$\frac{u}{g^{-1}(u)} = P(u) = \frac{u/2}{\sinh u/2}\, F(e^{u}+e^{-u}-2)\,, \tag{1}$$

where g^{-1} denotes the inverse power series of g.

Recently, a particular class of genera has come into prominence through the work of Landweber, Stong, Ochanine, Witten and others. These genera are characterized topologically by the property that $\varphi(M)$ vanishes if M is the total space of the complex projective bundle associated to an even-dimensional complex vector bundle over a closed oriented manifold, and numerically by the property that the power series $g'(x)^{-2}$ is a polynomial of degree ≤ 4, i.e., that

$$(2) \qquad g(x) = \int_0^x \frac{dt}{\sqrt{1-2\delta t^2+\varepsilon t^4}} \qquad \text{for some } \delta, \varepsilon \in R .$$

(The equivalence of these two definitions is due to Ochanine [5].) Since this is an elliptic integral, such φ are called <u>elliptic genera</u>. Landweber and Stong [4] discovered that there is a particular elliptic genus with values in the power series ring $R = \mathbb{Q}[[q]]$ satisfying:

(a) For $r \geq 1$ the coefficient of y^r in $F(y)$ belongs to $q^{2r-1}R$.

(This, or rather the weaker statement that the coefficient of y^r is divisible by q^{r+1} for $r \geq 2$, arises from a certain natural property of the above-mentioned KO-characteristic class $\mathcal{F}$ which we do not formulate here.) Based on numerical computations, they conjectured that condition (a) characterizes the genus in question up to a reparametrization (i.e., up to replacing q by $aq+bq^2+\ldots$ with $a \neq 0$) and that with a suitable choice of parameter one has

(b) $F(y)$ has coefficients in $\mathbb{Z}[[q]]$.

By what was said in (iii), this means that the genus takes on values in $\mathbb{Z}[[q]]$ for all Spin manifolds. These facts were proved by D. and G. Chudnovsky [2], whose formulas show that with a suitable normalization one also has

(c) The leading term of the coefficient of y^r in $F(y)$ for $r \geq 1$ is $-q^{2r-1}$, and

(d) The genus takes values in the subring $M_*^{\mathbb{Q}}(\Gamma_0(2)) \subset \mathbb{Q}[[q]]$ of modular forms on $\Gamma_0(2)$ with rational Fourier coefficients.

(We recall basic definitions about modular forms below.) In particular, the δ and ε of equation (2) are certain modular forms (of weights 2 and 4); since $M_*^{\mathbb{Q}}(\Gamma_0(2))$ is known to be the free polynomial algebra on δ and ε, it follows that the Landweber-Stong genus is universal for all elliptic genera. This universal, modular form-valued elliptic genus has been the object of considerable interest; it gives rise to new cohomology theories (the "elliptic cohomology" of Landweber, Stong, and Ravenel) and to connections with index theory, string theory, etc. The purpose of this note is to give elementary proofs of a variety of formulas for the power series g, P, and F associated to the Landweber-Stong genus (and in particular, easy proofs of the properties (a)-(d)). These proofs use ideas from the theory of elliptic functions and modular forms but we will prove everything we need from scratch.

THEOREM. <u>Let</u> $R = \mathbb{Q}[[q]]$. <u>Then the following five formulas define the same power series</u> $P(u) \in R[[u]]$:

$$(3) \qquad P(u) = 1 - \sum_{\substack{k>0 \\ 2|k}} \frac{G_k^*}{2^{k-2}(k-1)!} u^k ,$$

$$(4) \qquad P(u) = \exp\Big(\sum_{\substack{k>0 \\ 2|k}} \frac{2\widetilde{G}_k}{k!} u^k \Big) ,$$

(5) $$P(u) = \frac{u}{g^{-1}(u)}$$ with g given by (2) ,

(6) $$P(u) = \frac{u/2}{\sinh u/2} \prod_{n=1}^{\infty} \left[\frac{(1-q^n)^2}{(1-q^n e^u)(1-q^n e^{-u})}\right]^{(-1)^n} ,$$

(7) $$P(u) = \frac{u/2}{\sinh u/2} \cdot \left[1 - \sum_{r=1}^{\infty} a_r (e^u + e^{-u} - 2)^r \right] ,$$

where G_k^*, $\tilde{G}_k$, δ, ε and $a_r \in \mathbb{R}$ are defined by

$$G_k^* = G_k^*(q) = \frac{2^{k-1}-1}{2k} B_k + \sum_{n\geq 1} \Bigg(\sum_{\substack{d|n \\ 2\nmid d}} d^{k-1} \Bigg) q^n ,$$

$$\tilde{G}_k = \tilde{G}_k(q) = -\frac{1}{2k} B_k + \sum_{n\geq 1} \Bigg(\sum_{d|n} (-1)^{n/d} d^{k-1} \Bigg) q^n ,$$

$$\delta = -3 G_2^* = 3 \tilde{G}_2 = -\frac{1}{8} - 3 \sum_{n\geq 1} \Bigg(\sum_{\substack{d|n \\ 2\nmid d}} d \Bigg) q^n ,$$

$$\varepsilon = -\frac{1}{6}(G_4^* + 7\tilde{G}_4) = \sum_{n\geq 1} \Bigg(\sum_{\substack{d|n \\ 2\nmid n/d}} d^3 \Bigg) q^n ,$$

(8) $$a_r = \sum_{m\geq 1} \frac{q^{m(2r-1)}(1+q^{2m})}{(1-q^{2m})^{2r}} = \sum_{n\geq 1} \sum_{\substack{d|n \\ 2\nmid d}} \left[\binom{\frac{1}{2}(d-1)+r}{2r-1} + \binom{\frac{1}{2}(d-3)+r}{2r-1} \right] q^n$$

(here $B_2 = \frac{1}{6}$, $B_4 = -\frac{1}{30}$, ... are Bernoulli numbers and $\sum_{d|n}$ denotes a sum over positive divisors of n). The genus with characteristic power series $P(u)$ satisfies properties (a)-(d).

Each of the five formulas in the theorem describes some aspect of the genus with characteristic power series $P(u)$: (3) and (4) describe the genus in cohomology and make the modularity property (d) clear (since G_k^* and $\tilde{G}_k$ are the Fourier expansions of well-known Eisenstein series, as recalled below), (5) shows that the genus is elliptic, and (6) and (7) describe the genus in K-theory and (both) make the properties (a)-(c) evident. (To deduce (a) and (c) from (6) one has to split off the terms $n = 1$ and $n = 2$ from the infinite product.) Formula (6) was given by the Chudnovskys, but with a different proof. It has been generalized by Witten [6] to get other genera whose coefficients are modular forms, and in this form interpreted by him, using ideas from quantum field theory, as the equivariant index formula (Atiyah-Bott-Singer fixed point theorem) for a Dirac operator on the free loop space of a manifold. We shall return to these other genera at the end of the note.

Proof of the theorem. Consider meromorphic functions $\psi: \mathbb{C} \to \mathbb{C}$ satisfying

(9) $$\begin{cases} \psi(u+2\pi i) = -\psi(u) , \quad \psi(u+4\pi i\tau) = \psi(u) , \quad \psi(-u) = -\psi(u) , \\ \psi \text{ has poles only for } u \in L , \quad \psi(u) = \frac{1}{u} + O(1) \text{ as } u \to 0 , \end{cases}$$

where τ is in the complex upper half-plane and L denotes the lattice $\mathbb{Z}\cdot 4\pi i\tau + \mathbb{Z}\cdot 2\pi i$. Clearly there can be at most one such function, since the difference of any two would be holomorphic and doubly periodic, hence constant, hence zero because odd. We will give different constructions showing that ψ exists and equals $\frac{1}{u}P(u)$ for $P(u)$ given by any of the five equations (3)-(7), with $q = e^{2\pi i\tau}$.

First, define ψ by the rapidly convergent series

$$(10) \qquad \psi(u) = \sum_{m\in\mathbb{Z}} \frac{1}{2\sinh(\frac{u}{2} + 2\pi i m\tau)} = \sum_{m\in\mathbb{Z}} \frac{1}{q^m e^{u/2} - q^{-m}e^{-u/2}} .$$

The properties (9) are immediately checked. Combining the terms m and -m, we find

$$\psi(u) = \frac{1}{e^{u/2} - e^{-u/2}} - (e^{u/2} - e^{-u/2}) \sum_{m=1}^{\infty} \frac{q^m(1+q^{2m})}{(1-q^{2m}e^u)(1-q^{2m}e^{-u})}$$

or, setting $y = e^u + e^{-u} - 2$,

$$u\,\psi(u) = \frac{u/2}{\sinh u/2}\left(1 - \sum_{m=1}^{\infty} \frac{q^m(1+q^{2m})y}{(1-q^{2m})^2 - q^{2m}y}\right) .$$

Expanding the geometric series in y, we find the function $P(u)$ defined by (7), with a_r given by the first formula in (8). The second formula in (8) follows from the first by applying the binomial theorem; either one makes properties (a)-(c) evident.

Next, define $\psi(u)$ as $u^{-1}P(u)$ with $P(u)$ given by the product formula (6). Again it is easy to check that this function satisfies (9) and hence is the same as the one just considered. This proves (6) and gives a second proof of (a)-(c).

The Taylor expansions (3) and (4) of $P(u)$ and $\log P(u)$ are easily obtained from the above two constructions: the first construction gives

$$P(u) = \frac{u/2}{\sinh u/2} - \sum_{m=1}^{\infty}\left(\frac{q^m e^{u/2}}{1-q^{2m}e^u} - \frac{q^m e^{-u/2}}{1-q^{2m}e^{-u}}\right) = \frac{u/2}{\sinh u/2} - \sum_{\substack{m,d\geq 1\\ d\text{ odd}}} (e^{\frac{du}{2}} - e^{-\frac{du}{2}})\, q^{md} ,$$

which (on substituting the Taylor series of $\frac{u/2}{\sinh u/2}$ and $\sinh\frac{du}{2}$) is seen to be equivalent to (3), and the second gives

$$\log P(u) = \log\frac{u/2}{\sinh u/2} + \sum_{n=1}^{\infty} (-1)^n \sum_{d=1}^{\infty} (e^{du} + e^{-du} - 2)\,\frac{q^{nd}}{d} ,$$

which is similarly equivalent to (4) (the Taylor expansion of $\log\frac{u/2}{\sinh u/2}$ can be found by differentiation).

Finally, the "elliptic" property (5) also follows from the axiomatic characterization (9): the properties (9) imply that $\psi(u)^2$ and $\psi'(u)^2$ are even and invariant under translation by L and have poles at $u=0$ with leading terms u^{-2} and u^{-4} as their only singularities (mod L), so ψ'^2 must be a monic quadratic polynomial of ψ^2, i.e., $\psi'^2 = \psi^4 - 2\delta\psi^2 + \varepsilon$ for some $\delta, \varepsilon \in \mathbb{C}$, and then $\frac{1}{\psi(u)} = u + \ldots$ can be written as $g^{-1}(u)$ where $g(x) = x + \ldots$ is given by the elliptic integral (2). The expansions of δ and ε as functions of $q = e^{2\pi i\tau}$ can be obtained by comparing the coefficients of u^2 and u^4 in $P(u)$ obtained from formulas (3), (4), and (5).

It remains to check property (d), i.e., that the series $P(u)$ defined in the theorem has Taylor coefficients which are modular forms on $\Gamma_0(2)$. We recall that

$\Gamma_0(2)$ is the subgroup of $SL_2(\mathbb{Z})$ consisting of matrices $\binom{a\ b}{c\ d}$ with c even and that for $\Gamma = \Gamma_0(2)$ or $SL_2(\mathbb{Z})$ a modular form of weight k on Γ (k an integer, necessarily even and nonnegative) is a holomorphic function $f: \mathcal{H} \to \mathbb{C}$ ($\mathcal{H}$ = upper half-plane) satisfying $f(\frac{a\tau+b}{c\tau+d}) = (c\tau+d)^k f(\tau)$ for all $\tau \in \mathcal{H}$, $\binom{a\ b}{c\ d} \in \Gamma$ and having a Fourier expansion $\sum a(n)\,q^n$ with $a(n)$ of polynomial growth in n. The $\mathbb{C}$-vector space of such forms is denoted by $M_k(\Gamma)$, the $\mathbb{Q}$-vector space (of the same dimension) of forms with $a(n) \in \mathbb{Q}$ for all n is denoted by $M_k^{\mathbb{Q}}(\Gamma)$, and the graded ring $\bigoplus_k M_k^{\mathbb{Q}}(\Gamma)$ by $M_*^{\mathbb{Q}}(\Gamma)$. For $\Gamma = SL_2(\mathbb{Z})$ this ring is $\mathbb{Q}[G_4, G_6]$, where

$$(11)\qquad G_k = G_k(\tau) = -\frac{B_k}{2k} + \sum_{n=1}^{\infty}\Big(\sum_{d|n} d^{k-1}\Big)\,q^n \qquad (k>0,\ k \text{ even}).$$

For $k \geq 4$ the function $G_k(\tau)$ is modular of weight k because $\frac{2(2\pi i)^k}{(k-1)!}\,G_k$ is equal to the absolutely convergent <u>Eisenstein series</u> $\sum{}'(m\tau+n)^{-k}$ (summation over all pairs of integers m, n, not both zero). The function G_2 is "nearly" modular: it satisfies

$$(12)\qquad G_2\Big(\frac{a\tau+b}{c\tau+d}\Big) = (c\tau+d)^2\,G_2(\tau) + \frac{i}{4\pi}\,c(c\tau+d) \quad \text{for } \tau\in\mathcal{H},\ \binom{a\ b}{c\ d} \in SL_2(\mathbb{Z}).$$

From these facts and the easily checked identities

$$(13)\qquad G_k^*(\tau) = G_k(\tau) - 2^{k-1}G_k(2\tau), \qquad \widetilde{G}_k(\tau) = -G_k(\tau) + 2\,G_k(2\tau) \qquad (k \geq 2)$$

it follows that G_k^* and $\widetilde{G}_k$ are modular forms on $\Gamma_0(2)$ for all k (including k = 2). Therefore each of the formulas (3), (4), or (5) shows that P(u) has coefficients in $M_*^{\mathbb{Q}}(\Gamma_0(2))$; more precisely, the coefficient of u^k is in $M_k^{\mathbb{Q}}(\Gamma_0(2))$ for each $k \geq 0$. The modularity also follows from (10) because the expansion $\frac{1}{\sinh x} = \sum_{n\in\mathbb{Z}} \frac{(-1)^n}{x+\pi i n}$ gives

$$u\,\psi(u) = u\sum_{m\in\mathbb{Z}}\sum_{n\in\mathbb{Z}} \frac{(-1)^n}{u+4\pi i m\tau+2\pi i n} = 1 - \sum_{\substack{k>0\\ k\text{ even}}}\Big(\frac{u}{4\pi i}\Big)^k \sum_{m,n\in\mathbb{Z}}{}' \frac{(-1)^n}{(m\tau+n/2)^k}$$

with the inner sum clearly a modular form of weight k on $\Gamma_0(2)$ (here some care is needed with the conditionally convergent double series), or from the axiomatic characterization (9) by noting that for $\binom{a\ b}{c\ d} \in \Gamma_0(2)$ and ψ satisfying (9) the function $\tilde{\psi}(u) = (c\tau+d)\,\psi((c\tau+d)u)$ satisfies (9) with respect to $\tilde{\tau} = \frac{a\tau+b}{c\tau+d}$.

This completes the proof of the theorem. From a purely modular point of view, there are two surprising aspects of the formulas it contains. First of all, the space of modular forms $M_k(\Gamma_0(2))$ breaks up in a natural way as the direct sum of a space of Eisenstein series (forms whose Fourier coefficients a(n) are sums of powers of divisors of n with congruence conditions) and a space of "cusp forms" (forms satisfying $a(n) = O(n^{k/2})$). The former is spanned by G_k^* and $\widetilde{G}_k$ and hence has dimension 1 for k = 2 and 2 for k > 2 (the dimension of the full space M_k is k/2). It is quite remarkable that the coefficients of both P(u) and log P(u) belong to this tiny subspace. The other surprising fact is that, although the Eisenstein series $G_2^*, G_4^*, \ldots$ have rational Fourier coefficients and non-zero constant terms, there is a rational linear combination of $1, G_2^*, \ldots, G_{2r}^*$, namely a_r, which vanishes to order 2r − 1 (i.e. twice as far as one has any right to expect) and is monic with integral coefficients. This, and also the fact that the a_r have much smaller coefficients (i.e., that the G_k^* satisfy congruences to high moduli), are illustrated by the first values:

$$\begin{aligned}
G_2^* &= \tfrac{1}{24} + q + q^2 + 4q^3 + q^4 + 6q^5 + 4q^6 + 8q^7 + q^8 + 13q^9 + \dots \\
G_4^* &= -\tfrac{7}{240} + q + q^2 + 28q^3 + q^4 + 126q^5 + 28q^6 + 344q^7 + q^8 + 757q^9 + \dots \\
G_6^* &= \tfrac{31}{504} + q + q^2 + 244q^3 + q^4 + 3126q^5 + 244q^6 + 16808q^7 + q^8 + 59293q^9 + \dots \\
a_1 &= q + q^2 + 4q^3 + q^4 + 6q^5 + 4q^6 + 8q^7 + q^8 + 13q^9 + \dots \\
a_2 &= q^3 + 5q^5 + q^6 + 14q^7 + 31q^9 + \dots \\
a_3 &= q^5 + 7q^7 + 27q^9 + \dots
\end{aligned}$$

We now turn to the other genera introduced by Witten. The same formal power series calculation as that which showed the equivalence of (4) and (6) gives

$$(14) \qquad \frac{u/2}{\sinh u/2} \prod_{n=1}^{\infty} \frac{(1-q^n)^2}{(1-q^n e^u)(1-q^n e^{-u})} = \exp\Big(\sum_{\substack{k>0 \\ 2|k}} \frac{2}{k!} G_k u^k \Big)$$

where G_k is defined by (11). Call this power series $P_W(u)$ and the associated genus φ_W (W for "Witten" or "Weierstrass"; terminology due to Peter Landweber). Let us compare this genus and power series with those of Landweber-Stong:

- The left side of (14) shows that the KO-theory characteristic power series of φ_W can be written in the form

$$\prod_{n=1}^{\infty} \Big[1 - \frac{q^n}{(1-q^n)^2} y \Big]^{-1} = \sum_{r=0}^{\infty} b_r y^r \qquad (y = e^u - 2 + e^{-u})$$

where b_r belongs to $\mathbb{Z}[[q]]$ and has leading coefficient q^r. Thus the analogues of properties (a)-(c) hold for the Witten genus, and $\varphi_W(M)$ belongs to $\mathbb{Z}[[q]]$ if M is a Spin manifold. The product in (14) is simpler than that in (6). It can be neatly expressed by saying that the associated KO-valued characteristic class $\mathcal{F}_W(\xi)$ is a certain tensor product of sums of symmetric powers of ξ, and in this form has a natural interpretation as an index formula for a Dirac-like operator on the free loop space of M; the corresponding expression for the Landweber-Stong genus also has an interpretation as an index formula for an operator, but this time one which has no finite-dimensional version. (For all this, see Witten's paper [6].)

- The right side of (14) shows that $\varphi_W(M)$ is a modular form on $SL_2(\mathbb{Z})$ if the first Pontryagin class of M vanishes rationally, since then G_2 drops out of the characteristic class $\mathcal{F}_W(TM)$. In this case $\varphi_W(M)$ is simpler than $\varphi_{LS}(M)$ because it is a modular form on the full modular group rather than a congruence subgroup. On the other hand, if $p_1(M)$ does not vanish then $\varphi_W(M)$ is not a modular form at all, but belongs instead to the larger ring

$$\hat{M}_* = \mathbb{Q}[G_2, G_4, G_6] \supset M_* = M_*^{\mathbb{Q}}(SL_2(\mathbb{Z})) = \mathbb{Q}[G_4, G_6] .$$

This may not be all bad: the ring $\hat{M}_*$ is also studied in the theory of modular forms and is in many respects nearly as good as M_* (the elements of $\hat{M}_*$ are "almost modular" by (12), and the mod p reductions of $\hat{M}_* \cap \mathbb{Z}[[q]]$ and $M_* \cap \mathbb{Z}[[q]]$ agree for all primes p). In one respect it is even better: it is closed under the differentiation operator $D = q\frac{d}{dq} = \frac{1}{2\pi i}\frac{d}{d\tau}$ (specifically, $DG_2 = \frac{5}{6}G_4 - 2G_2^2$, $DG_4 = \frac{7}{10}G_6 - 8G_2G_4$, $DG_6 = \frac{400}{7}G_4^2 - 12G_2G_6$, and more generally $Df + 2kG_2f \in M_{k+2}$ for $f \in M_k$).

- There is no analogue of the additive formulae (3) and (7) or of the elliptic property (5), so φ_W is not an elliptic genus and does not have a simple description in cohomology. This is because the axiomatic characterization (9) no longer applies: the function $u^{-1}P_W(u)$ is nearly, but not quite, doubly periodic. (It changes by a factor $-q^{1/2}e^u$ under $u \to u + 2\pi i\tau$. The function $\frac{d^2}{du^2}\log u^{-1}P_W(u)$ is periodic, and in fact equals $2G_2$ minus the Weierstrass $\wp$-function for the lattice $\mathbb{Z}\cdot 2\pi i\tau + \mathbb{Z}\cdot 2\pi i$.) This is closely related to the non-modularity of the coefficients of $P_W(u)$ as functions of τ (cf. comments at the end of the note).

Witten discusses one other genus, the one associated to the signature operator. Let $\mathcal{L}(\xi)$ and $\mathcal{L}_n(\xi)$ $(n \geq 1)$ be the characteristic classes with characteristic power series $\frac{u}{\tanh u}$ and $\frac{1+q^ne^{2u}}{1-q^ne^{2u}}$, where q is a parameter. The G-signature theorem of Atiyah and Singer says that for a smooth finite-dimensional manifold X with S^1-action, the equivariant signature, an element of the representation ring $R(S^1)\otimes\mathbb{C} \simeq \mathbb{C}[q,q^{-1}]$, is given by

$$(15) \qquad \langle \mathcal{L}(TM)\prod_{n=1}^{\infty}\mathcal{L}_n(\nu_n),\ [M]\rangle ,$$

where $M = X^{S^1}$ is the fixed point set and ν_n the subbundle of the normal bundle of M in X on which S^1 acts via $\begin{pmatrix}\cos n\theta & \sin n\theta\\ -\sin n\theta & \cos n\theta\end{pmatrix}$. (We are being very brief and a little imprecise here.) This equivariant signature is constant, either (cf. [1])

(i) because it is defined in terms of the action of S^1 on the middle cohomology of X and this action is trivial since S^1 is connected, or

(ii) because it is an element of $\mathbb{C}[q,q^{-1}]$ which is regular at both $q=0$ and $q=\infty$.

Witten's idea was to apply (15) to the case when X is the free loop space of a smooth manifold M; then M is the fixed point set and each ν_n is finite-dimensional (and in fact isomorphic to $TM\otimes\mathbb{C}$), so that the formula makes sense even though X itself is infinite-dimensional. On the other hand, neither (i) nor (ii) applies and we get a non-trivial power series in q. Modifying $\mathcal{L}_n$ by dividing its defining characteristic power series by its value at $u=0$ (i.e., looking at the associated stable class), and replacing u by u/2, we find that this power series is $\left(2^{\frac{1}{2}}\prod_n\frac{1+q^n}{1-q^n}\right)^{\dim M}\varphi_S(M)$, where φ_S is the $\mathbb{Q}[[q]]$-valued genus with characteristic power series

$$(16) \qquad P_S(u) = \frac{u/2}{\tanh u/2}\prod_{n=1}^{\infty}\left(\frac{1+q^ne^u}{1-q^ne^u}\cdot\frac{1+q^ne^{-u}}{1-q^ne^{-u}}\right)\Big/\left(\frac{1+q^n}{1-q^n}\right)^2 .$$

For this genus, analogues of all five formulas in the theorem about P(u) hold. The analogue of (6) is (16) itself, and the analogues of (3), (4), (5), and (7) are:

(17) $$P_S(u) = 1 - \sum_{\substack{k>0\\ 2|k}} \frac{2\,\widetilde{G}_k}{(k-1)!}\,u^k ,$$

(18) $$P_S(u) = \exp\Big(\sum_{\substack{k>0\\ 2|k}} \frac{4\,G_k^*}{k!}\,u^k\Big) ,$$

(19) $$P_S(u) = u\,/\,g_S^{-1}(u), \quad \text{where} \quad g_S(x) = \int_0^x (1-2\delta_S t^2 + \varepsilon_S t^4)^{-\frac12}\,dt \quad \text{with}$$

$$\delta_S = \frac14 + 6\sum_{n\ge1}\Big(\sum_{\substack{d|n\\ 2\nmid d}} d\Big)q^n, \qquad \varepsilon_S = \frac1{16} + \sum_{n\ge1}\Big(\sum_{d|n}(-1)^d d^3\Big)q^n ,$$

(20) $$P_S(u) = \frac{u/2}{\tanh u/2}\Big(1 - 2\sum_{r\ge1} c_r\,(e^u+e^{-u}-2)^r\Big) \quad \text{with}$$

$$c_r = \sum_{m\ge1}\frac{(-1)^m q^{mr}}{(1-q^m)^{2r}} = \sum_{n\ge1}\Big[\sum_{d|n}(-1)^{n/d}\binom{r+d-1}{2r-1}\Big]q^n \in q^r\,\mathbb{Z}[[q]] .$$

Indeed, (18) is obtained directly from (16) by logarithmic differentiation (like the proof of (14) or of the equality of (4) and (6)). But from equation (13) and the modularity property $G_k(\frac{-1}{\tau}) = \tau^k G_k(\tau)$ (respectively (12) for $k=2$) it follows that

(21) $$G_k^*(\tfrac{-1}{2\tau}) = 2^{k-1}\,\tau^k\,\widetilde{G}_k(\tau) , \qquad \widetilde{G}_k(\tfrac{-1}{2\tau}) = 2\,\tau^k\,G_k^*(\tau) ,$$

so (writing $P(\tau;u)$ instead of $P(u)$ for the Landweber-Stong genus to emphasize the dependence on τ, and similarly for P_S) equation (18) says

(22) $$P_S(\tau;u) = P(\tfrac{-1}{2\tau};\tfrac{u}{\tau}) .$$

In other words, P and P_S are the same function, but expanded at the two cusps 0 and ∞ of $H/\Gamma_0(2)$ (which are interchanged by $\tau \to -1/2\tau$). Substituting formulas (3) and (5) into (22) and using (21) again gives (17) and (19). Finally, equation (22) leads to an analogue of the property (9) and hence to an analogue of (10), namely

$$u^{-1}P_S(u) = \sum_{m\in\mathbb{Z}}\frac{(-1)^m}{2\tanh(u/2+\pi i m\tau)} = \frac{1/2}{\tanh u/2} - \sum_{m\ge1}\frac{(-1)^m\,(e^u-e^{-u})}{q^m+q^{-m}-e^u-e^{-u}} ,$$

from which (20) easily follows. Equation (20) expresses $P_S(u)$ as $\frac{u/2}{\tanh u/2}G(e^u+e^{-u}-2)$ with $G(y) \in \mathbb{Z}[[q, qy]]$; such a formula has an interpretation like the one given for equation (1) in (iii) of the introduction, but using a different KO_*-fundamental class (the one associated to the signature operator).

We make a final remark. Throughout this note there has been an interplay between the modularity properties of the various functions with respect to the variable τ and their elliptic properties with respect to the variable u. Functions of two variables having this dual modular/elliptic nature are called Jacobi forms. More precisely, a Jacobi form of weight k and index m is a function $\phi\colon H\times\mathbb{C} \to \mathbb{C}$ satisfying

$$\phi(\frac{a\tau+b}{c\tau+d},\frac{z}{c\tau+d}) = (c\tau+d)^k\,e^{2\pi i m c z^2/(c\tau+d)}\,\phi(\tau,z)$$

for $\binom{a\ b}{c\ d}$ in $SL_2(\mathbb{Z})$ or a congruence subgroup and

$$\phi(\tau, z+\lambda\tau+\mu) = e^{-2\pi i m(\lambda^2\tau+2\lambda z)}\,\phi(\tau,z)$$

for all (λ,μ) in $\mathbb{Z}^2$ or a sublattice of finite index. (The theory of such forms was developed in [3].) The most important examples of Jacobi forms are theta series. Using the Jacobi triple product identity, we find that $u^{-1}P_W(u)$ can be expressed as a quotient of theta-series,

$$u^{-1}P_W(u) = \Big(\sum_{n>0}\left(\tfrac{-4}{n}\right) n\, q^{n^2/8}\Big)\Big/\Big(\sum_{n\in\mathbb{Z}}\left(\tfrac{-4}{n}\right) q^{n^2/8}e^{nu/2}\Big)$$

(here $\left(\frac{-4}{n}\right)$ equals 0 for n even, $(-1)^{\frac{1}{2}(n-1)}$ for n odd), and is therefore a Jacobi form with respect to τ and $z=\frac{u}{2\pi i}$, of weight -1 and index $-\frac{1}{2}$. (It is because the index is non-zero that $u^{-1}P_W(u)$ is not quite elliptic in u and that its Taylor coefficients are not quite modular forms in τ.) The other two characteristic power series we have been considering are related to P_W by

$$P(u) = P_W(2\tau;u)^2/P_W(\tau;u), \qquad P_S(u) = P_W(\tau;u)^2/P_W(2\tau;2u),$$

so $u^{-1}P(u)$ and $u^{-1}P_S(u)$ are also Jacobi forms of weight -1, but of index 0. It is interesting to note that in all of the recent occurrences of modular forms in algebraic topology, string theory, and the theory of Kac-Moody algebras, it is in fact Jacobi forms which are entering.

REFERENCES

1. M.F. Atiyah and F. Hirzebruch: Spin-manifolds and group actions. In: Essays on Topology and Related Topics (Mémoires dédiés à Georges de Rham), pp. 18-26. Springer, New York 1970.

2. D.V. Chudnovsky and G.V. Chudnovsky: Elliptic modular functions and elliptic genera. To appear in Topology.

3. M. Eichler and D. Zagier: The Theory of Jacobi Forms. Progress in Math. 55, Birkhäuser, Boston-Basel-Stuttgart 1985.

4. P. Landweber and R.E. Stong: Circle actions on Spin manifolds and characteristic numbers. To appear in Topology.

5. S. Ochanine: Sur les genres multiplicatifs définis par des intégrales elliptiques. To appear in Topology 26 (1987) 143-151

6. E. Witten: The index of the Dirac operator in loop space. To appear in this volume.

Vol. 1173: H. Delfs, M. Knebusch, Locally Semialgebraic Spaces. XVI, 329 pages. 1985.

Vol. 1174: Categories in Continuum Physics, Buffalo 1982. Seminar. Edited by F.W. Lawvere and S.H. Schanuel. V, 126 pages. 1986.

Vol. 1175: K. Mathiak, Valuations of Skew Fields and Projective Hjelmslev Spaces. VII, 116 pages. 1986.

Vol. 1176: R.R. Bruner, J.P. May, J.E. McClure, M. Steinberger, H_∞ Ring Spectra and their Applications. VII, 388 pages. 1986.

Vol. 1177: Representation Theory I. Finite Dimensional Algebras. Proceedings, 1984. Edited by V. Dlab, P. Gabriel and G. Michler. XV, 340 pages. 1986.

Vol. 1178: Representation Theory II. Groups and Orders. Proceedings, 1984. Edited by V. Dlab, P. Gabriel and G. Michler. XV, 370 pages. 1986.

Vol. 1179: Shi J.-Y. The Kazhdan-Lusztig Cells in Certain Affine Weyl Groups. X, 307 pages. 1986.

Vol. 1180: R. Carmona, H. Kesten, J.B. Walsh, École d'Été de Probabilités de Saint-Flour XIV – 1984. Édité par P.L. Hennequin. X, 438 pages. 1986.

Vol. 1181: Buildings and the Geometry of Diagrams, Como 1984. Seminar. Edited by L. Rosati. VII, 277 pages. 1986.

Vol. 1182: S. Shelah, Around Classification Theory of Models. VII, 279 pages. 1986.

Vol. 1183: Algebra, Algebraic Topology and their Interactions. Proceedings, 1983. Edited by J.-E. Roos. XI, 396 pages. 1986.

Vol. 1184: W. Arendt, A. Grabosch, G. Greiner, U. Groh, H.P. Lotz, U. Moustakas, R. Nagel, F. Neubrander, U. Schlotterbeck, One-parameter Semigroups of Positive Operators. Edited by R. Nagel. X, 460 pages. 1986.

Vol. 1185: Group Theory, Beijing 1984. Proceedings. Edited by Tuan H.F. V, 403 pages. 1986.

Vol. 1186: Lyapunov Exponents. Proceedings, 1984. Edited by L. Arnold and V. Wihstutz. VI, 374 pages. 1986.

Vol. 1187: Y. Diers, Categories of Boolean Sheaves of Simple Algebras. VI, 168 pages. 1986.

Vol. 1188: Fonctions de Plusieurs Variables Complexes V. Séminaire, 1979–85. Edité par François Norguet. VI, 306 pages. 1986.

Vol. 1189: J. Lukeš, J. Malý, L. Zajíček, Fine Topology Methods in Real Analysis and Potential Theory. X, 472 pages. 1986.

Vol. 1190: Optimization and Related Fields. Proceedings, 1984. Edited by R. Conti, E. De Giorgi and F. Giannessi. VIII, 419 pages. 1986.

Vol. 1191: A.R. Its, V.Yu. Novokshenov, The Isomonodromic Deformation Method in the Theory of Painlevé Equations. IV, 313 pages. 1986.

Vol. 1192: Equadiff 6. Proceedings, 1985. Edited by J. Vosmansky and M. Zlámal. XXIII, 404 pages. 1986.

Vol. 1193: Geometrical and Statistical Aspects of Probability in Banach Spaces. Proceedings, 1985. Edited by X. Fernique, B. Heinkel, M.B. Marcus and P.A. Meyer. IV, 128 pages. 1986.

Vol. 1194: Complex Analysis and Algebraic Geometry. Proceedings, 1985. Edited by H. Grauert. VI, 235 pages. 1986.

Vol. 1195: J.M. Barbosa, A.G. Colares, Minimal Surfaces in $\mathbb{R}^3$. X, 124 pages. 1986.

Vol. 1196: E. Casas-Alvero, S. Xambó-Descamps, The Enumerative Theory of Conics after Halphen. IX, 130 pages. 1986.

Vol. 1197: Ring Theory. Proceedings, 1985. Edited by F.M.J. van Oystaeyen. V, 231 pages. 1986.

Vol. 1198: Séminaire d'Analyse, P. Lelong – P. Dolbeault – H. Skoda. Seminar 1983/84. X, 260 pages. 1986.

Vol. 1199: Analytic Theory of Continued Fractions II. Proceedings, 1985. Edited by W.J. Thron. VI, 299 pages. 1986.

Vol. 1200: V.D. Milman, G. Schechtman, Asymptotic Theory of Finite Dimensional Normed Spaces. With an Appendix by M. Gromov. VIII, 156 pages. 1986.

Vol. 1201: Curvature and Topology of Riemannian Manifolds. Proceedings, 1985. Edited by K. Shiohama, T. Sakai and T. Sunada. VII, 336 pages. 1986.

Vol. 1202: A. Dür, Möbius Functions, Incidence Algebras and Power Series Representations. XI, 134 pages. 1986.

Vol. 1203: Stochastic Processes and Their Applications. Proceedings, 1985. Edited by K. Itô and T. Hida. VI, 222 pages. 1986.

Vol. 1204: Séminaire de Probabilités XX, 1984/85. Proceedings. Edité par J. Azéma et M. Yor. V, 639 pages. 1986.

Vol. 1205: B.Z. Moroz, Analytic Arithmetic in Algebraic Number Fields. VII, 177 pages. 1986.

Vol. 1206: Probability and Analysis, Varenna (Como) 1985. Seminar. Edited by G. Letta and M. Pratelli. VIII, 280 pages. 1986.

Vol. 1207: P.H. Bérard, Spectral Geometry: Direct and Inverse Problems. With an Appendix by G. Besson. XIII, 272 pages. 1986.

Vol. 1208: S. Kaijser, J.W. Pelletier, Interpolation Functors and Duality. IV, 167 pages. 1986.

Vol. 1209: Differential Geometry, Peñíscola 1985. Proceedings. Edited by A.M. Naveira, A. Ferrández and F. Mascaró. VIII, 306 pages. 1986.

Vol. 1210: Probability Measures on Groups VIII. Proceedings, 1985. Edited by H. Heyer. X, 386 pages. 1986.

Vol. 1211: M.B. Sevryuk, Reversible Systems. V, 319 pages. 1986.

Vol. 1212: Stochastic Spatial Processes. Proceedings, 1984. Edited by P. Tautu. VIII, 311 pages. 1986.

Vol. 1213: L.G. Lewis, Jr., J.P. May, M. Steinberger, Equivariant Stable Homotopy Theory. IX, 538 pages. 1986.

Vol. 1214: Global Analysis – Studies and Applications II. Edited by Yu.G. Borisovich and Yu.E. Gliklikh. V, 275 pages. 1986.

Vol. 1215: Lectures in Probability and Statistics. Edited by G. del Pino and R. Rebolledo. V, 491 pages. 1986.

Vol. 1216: J. Kogan, Bifurcation of Extremals in Optimal Control. VIII, 106 pages. 1986.

Vol. 1217: Transformation Groups. Proceedings, 1985. Edited by S. Jackowski and K. Pawalowski. X, 396 pages. 1986.

Vol. 1218: Schrödinger Operators, Aarhus 1985. Seminar. Edited by E. Balslev. V, 222 pages. 1986.

Vol. 1219: R. Weissauer, Stabile Modulformen und Eisensteinreihen. III, 147 Seiten. 1986.

Vol. 1220: Séminaire d'Algèbre Paul Dubreil et Marie-Paule Malliavin. Proceedings, 1985. Edité par M.-P. Malliavin. IV, 200 pages. 1986.

Vol. 1221: Probability and Banach Spaces. Proceedings, 1985. Edited by J. Bastero and M. San Miguel. XI, 222 pages. 1986.

Vol. 1222: A. Katok, J.-M. Strelcyn, with the collaboration of F. Ledrappier and F. Przytycki, Invariant Manifolds, Entropy and Billiards; Smooth Maps with Singularities. VIII, 283 pages. 1986.

Vol. 1223: Differential Equations in Banach Spaces. Proceedings, 1985. Edited by A. Favini and E. Obrecht. VIII, 299 pages. 1986.

Vol. 1224: Nonlinear Diffusion Problems, Montecatini Terme 1985. Seminar. Edited by A. Fasano and M. Primicerio. VIII, 188 pages. 1986.

Vol. 1225: Inverse Problems, Montecatini Terme 1986. Seminar. Edited by G. Talenti. VIII, 204 pages. 1986.

Vol. 1226: A. Buium, Differential Function Fields and Moduli of Algebraic Varieties. IX, 146 pages. 1986.

Vol. 1227: H. Helson, The Spectral Theorem. VI, 104 pages. 1986.

Vol. 1228: Multigrid Methods II. Proceedings, 1985. Edited by W. Hackbusch and U. Trottenberg. VI, 336 pages. 1986.

Vol. 1229: O. Bratteli, Derivations, Dissipations and Group Actions on C*-algebras. IV, 277 pages. 1986.

Vol. 1230: Numerical Analysis. Proceedings, 1984. Edited by J.-P. Hennart. X, 234 pages. 1986.

Vol. 1231: E.-U. Gekeler, Drinfeld Modular Curves. XIV, 107 pages. 1986.

Vol. 1232: P.C. Schuur, Asymptotic Analysis of Soliton Problems. VIII, 180 pages. 1986.

Vol. 1233: Stability Problems for Stochastic Models. Proceedings, 1985. Edited by V.V. Kalashnikov, B. Penkov and V.M. Zolotarev. VI, 223 pages. 1986.

Vol. 1234: Combinatoire énumérative. Proceedings, 1985. Edité par G. Labelle et P. Leroux. XIV, 387 pages. 1986.

Vol. 1235: Séminaire de Théorie du Potentiel, Paris, No. 8. Directeurs: M. Brelot, G. Choquet et J. Deny. Rédacteurs: F. Hirsch et G. Mokobodzki. III, 209 pages. 1987.

Vol. 1236: Stochastic Partial Differential Equations and Applications. Proceedings, 1985. Edited by G. Da Prato and L. Tubaro. V, 257 pages. 1987.

Vol. 1237: Rational Approximation and its Applications in Mathematics and Physics. Proceedings, 1985. Edited by J. Gilewicz, M. Pindor and W. Siemaszko. XII, 350 pages. 1987.

Vol. 1238: M. Holz, K.-P. Podewski and K. Steffens, Injective Choice Functions. VI, 183 pages. 1987.

Vol. 1239: P. Vojta, Diophantine Approximations and Value Distribution Theory. X, 132 pages. 1987.

Vol. 1240: Number Theory, New York 1984–85. Seminar. Edited by D.V. Chudnovsky, G.V. Chudnovsky, H. Cohn and M.B. Nathanson. V, 324 pages. 1987.

Vol. 1241: L. Gårding, Singularities in Linear Wave Propagation. III, 125 pages. 1987.

Vol. 1242: Functional Analysis II, with Contributions by J. Hoffmann-Jørgensen et al. Edited by S. Kurepa, H. Kraljević and D. Butković. VII, 432 pages. 1987.

Vol. 1243: Non Commutative Harmonic Analysis and Lie Groups. Proceedings, 1985. Edited by J. Carmona, P. Delorme and M. Vergne. V, 309 pages. 1987.

Vol. 1244: W. Müller, Manifolds with Cusps of Rank One. XI, 158 pages. 1987.

Vol. 1245: S. Rallis, L-Functions and the Oscillator Representation. XVI, 239 pages. 1987.

Vol. 1246: Hodge Theory. Proceedings, 1985. Edited by E. Cattani, F. Guillén, A. Kaplan and F. Puerta. VII, 175 pages. 1987.

Vol. 1247: Séminaire de Probabilités XXI. Proceedings. Edité par J. Azéma, P.A. Meyer et M. Yor. IV, 579 pages. 1987.

Vol. 1248: Nonlinear Semigroups, Partial Differential Equations and Attractors. Proceedings, 1985. Edited by T.L. Gill and W.W. Zachary. IX, 185 pages. 1987.

Vol. 1249: I. van den Berg, Nonstandard Asymptotic Analysis. IX, 187 pages. 1987.

Vol. 1250: Stochastic Processes – Mathematics and Physics II. Proceedings 1985. Edited by S. Albeverio, Ph. Blanchard and L. Streit. VI, 359 pages. 1987.

Vol. 1251: Differential Geometric Methods in Mathematical Physics. Proceedings, 1985. Edited by P.L. García and A. Pérez-Rendón. VII, 300 pages. 1987.

Vol. 1252: T. Kaise, Représentations de Weil et GL_2 Algèbres de division et GL_n. VII, 203 pages. 1987.

Vol. 1253: J. Fischer, An Approach to the Selberg Trace Formula via the Selberg Zeta-Function. III, 184 pages. 1987.

Vol. 1254: S. Gelbart, I. Piatetski-Shapiro, S. Rallis. Explicit Constructions of Automorphic L-Functions. VI, 152 pages. 1987.

Vol. 1255: Differential Geometry and Differential Equations. Proceedings, 1985. Edited by C. Gu, M. Berger and R.L. Bryant. XII, 243 pages. 1987.

Vol. 1256: Pseudo-Differential Operators. Proceedings, 1986. Edited by H.O. Cordes, B. Gramsch and H. Widom. X, 479 pages. 1987.

Vol. 1257: X. Wang, On the C*-Algebras of Foliations in the Plane. V, 165 pages. 1987.

Vol. 1258: J. Weidmann, Spectral Theory of Ordinary Differential Operators. VI, 303 pages. 1987.

Vol. 1259: F. Cano Torres, Desingularization Strategies for Three-Dimensional Vector Fields. IX, 189 pages. 1987.

Vol. 1260: N.H. Pavel, Nonlinear Evolution Operators and Semi-groups. VI, 285 pages. 1987.

Vol. 1261: H. Abels, Finite Presentability of S-Arithmetic Groups. Compact Presentability of Solvable Groups. VI, 178 pages. 1987.

Vol. 1262: E. Hlawka (Hrsg.), Zahlentheoretische Analysis II. Seminar, 1984–86. V, 158 Seiten. 1987.

Vol. 1263: V.L. Hansen (Ed.), Differential Geometry. Proceedings, 1985. XI, 288 pages. 1987.

Vol. 1264: Wu Wen-tsün, Rational Homotopy Type. VIII, 219 pages. 1987.

Vol. 1265: W. Van Assche, Asymptotics for Orthogonal Polynomials. VI, 201 pages. 1987.

Vol. 1266: F. Ghione, C. Peskine, E. Sernesi (Eds.), Space Curves. Proceedings, 1985. VI, 272 pages. 1987.

Vol. 1267: J. Lindenstrauss, V.D. Milman (Eds.), Geometrical Aspects of Functional Analysis. Seminar. VII, 212 pages. 1987.

Vol. 1268: S.G. Krantz (Ed.), Complex Analysis. Seminar, 1986. VII, 195 pages. 1987.

Vol. 1269: M. Shiota, Nash Manifolds. VI, 223 pages. 1987.

Vol. 1270: C. Carasso, P.-A. Raviart, D. Serre (Eds.), Nonlinear Hyperbolic Problems. Proceedings, 1986. XV, 341 pages. 1987.

Vol. 1271: A.M. Cohen, W.H. Hesselink, W.L.J. van der Kallen, J.R. Strooker (Eds.), Algebraic Groups Utrecht 1986. Proceedings. XII, 284 pages. 1987.

Vol. 1272: M.S. Livšic, L.L. Waksman, Commuting Nonselfadjoint Operators in Hilbert Space. III, 115 pages. 1987.

Vol. 1273: G.-M. Greuel, G. Trautmann (Eds.), Singularities, Representation of Algebras, and Vector Bundles. Proceedings, 1985. XIV, 383 pages. 1987.

Vol. 1274: N. C. Phillips, Equivariant K-Theory and Freeness of Group Actions on C*-Algebras. VIII, 371 pages. 1987.

Vol. 1275: C.A. Berenstein (Ed.), Complex Analysis I. Proceedings, 1985–86. XV, 331 pages. 1987.

Vol. 1276: C.A. Berenstein (Ed.), Complex Analysis II. Proceedings, 1985–86. IX, 320 pages. 1987.

Vol. 1277: C.A. Berenstein (Ed.), Complex Analysis III. Proceedings, 1985–86. X, 350 pages. 1987.

Vol. 1278: S.S. Koh (Ed.), Invariant Theory. Proceedings, 1985. V, 102 pages. 1987.

Vol. 1279: D. Ieşan, Saint-Venant's Problem. VIII, 162 Seiten. 1987.

Vol. 1280: E. Neher, Jordan Triple Systems by the Grid Approach. XII, 193 pages. 1987.

Vol. 1281: O.H. Kegel, F. Menegazzo, G. Zacher (Eds.), Group Theory. Proceedings, 1986. VII, 179 pages. 1987.

Vol. 1282: D.E. Handelman, Positive Polynomials, Convex Integral Polytopes, and a Random Walk Problem. XI, 136 pages. 1987.

Vol. 1283: S. Mardešić, J. Segal (Eds.), Geometric Topology and Shape Theory. Proceedings, 1986. V, 261 pages. 1987.

Vol. 1284: B.H. Matzat, Konstruktive Galoistheorie. X, 286 pages. 1987.

Vol. 1285: I.W. Knowles, Y. Saitō (Eds.), Differential Equations and Mathematical Physics. Proceedings, 1986. XVI, 499 pages. 1987.

Vol. 1286: H.R. Miller, D.C. Ravenel (Eds.), Algebraic Topology. Proceedings, 1986. VII, 341 pages. 1987.

Vol. 1287: E.B. Saff (Ed.), Approximation Theory, Tampa. Proceedings, 1985–1986. V, 228 pages. 1987.

Vol. 1288: Yu. L. Rodin, Generalized Analytic Functions on Riemann Surfaces. V, 128 pages, 1987.

Vol. 1289: Yu. I. Manin (Ed.), K-Theory, Arithmetic and Geometry. Seminar, 1984–1986. V, 399 pages. 1987.